Grade K

Getting Ready for the **New Jersey State Assessment**

INCLUDES

- New Jersey Student Learning Standards Practice in State Assessment format
- Beginning-, Middle-, and End-of-Year Benchmark Tests with Performance Tasks
- Year-End Performance Assessment Task

Printed in the U.S.A.

ISBN 978-1-328-96068-9

1 2 3 4 5 6 7 8 9 10 0982 22 21 20 19 18 17

4500667468 B C D E F G

Contents

New Jersey State Assessment Formats

The mathematics assessment for New Jersey contains item types beyond the traditional multiple-choice format, which allows for a more robust assessment of children's understanding of concepts.

The mathematics assessment for New Jersey will be administered via computers; and *Getting Ready for the New Jersey State Assessment* presents items in formats similar to what you will see on the tests. The following information is provided to help familiarize you with these different types of items. Each item type is identified on pages (vi–vii). The examples will introduce you to the item types.

The following explanations are provided to guide you in answering the questions. This page (v) describes the most common item types. You may find other types on some tests.

Example 1 Choose groups that have more than a given group.

More Than One Correct Choice

This type of item will ask you to choose all of something. When the item asks you to find all, look for more than one correct choice. Carefully look at each choice and mark it if it is a correct answer. Some items will ask you to "choose all." You may need to fill in bubbles instead of circling.

Example 2 Choose a number for a given group.

Choose From a List

Sometimes when you take a test on a computer, you will have to select a word or number from a drop-down list. The *Getting Ready for the New Jersey State Assessment* tests show a list and ask you to choose the correct answer. Make your choice by circling the correct answer. There will only be one choice that is correct.

Example 3 Match numbers.

Matching

Some items will ask you to match numbers or objects that are the same or are related in some way. The directions will specify what you should match. There will be dots to guide you in drawing lines. The matching may be between columns or rows.

Item Types:

Example 1

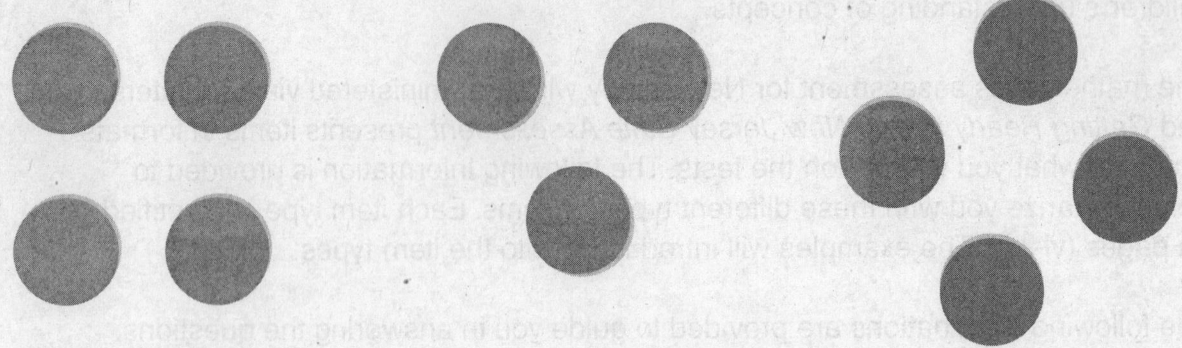

Example 2

DIRECTIONS 1. Choose all the groups that show more than 3 counters. 2. How many counters are there? Circle the number.

Example 3

0 3 5

3 0 5

DIRECTIONS **3.** Draw lines to match the same numbers.

Name ______________________

Practice Test
K.CC.A.1
Know number names and the count sequence.

21	22	23	24	25	26	27	28	29	30
31	32	33	34	35	36	37	38	39	40

50 ○ 60 ○ 70 ○ 80 ○

11	12	13	14	15	16	17	18	19	20
21	22	23	24	25	26	27	28	29	30

DIRECTIONS 1. Begin with 21. Point to each number as you count. Draw a line under the number to complete the counting order. 2. Count the crayons by tens. Mark under the number that shows how many. 3. Circle the numbers that complete each row of 10.

GO ON

Name ____________________

○ 40 ○ 50 ○ 60 ○ 70

5

83 84 85 86 87 | 88 / 90 | 89

6

94 95 96 97 98 99 | 90 / 100

DIRECTIONS 4. Count the straws by tens. Mark the number that shows how many. 5. Point to each number as you count. Circle the number to complete the counting order. 6. Point to each number as you count. Circle the number to complete the counting order.

Name ________________________________

Practice Test
K.CC.A.2
Know number names and the count sequence.

1

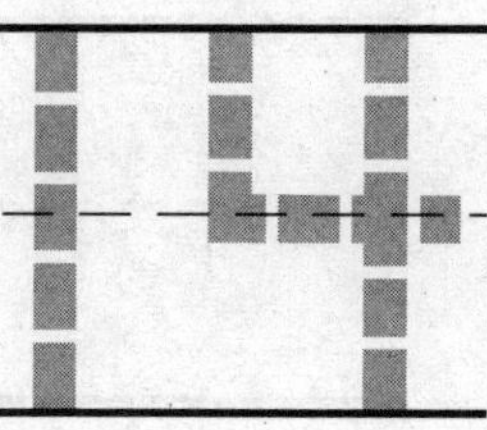

18

16

2

17 18 19 ○

10 11 12 ○

20 14 16 ○

3

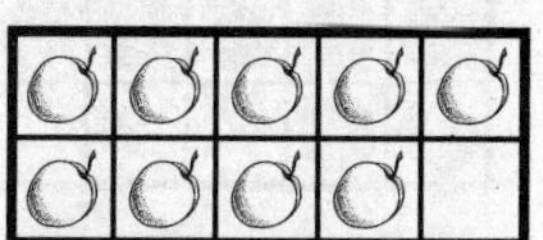

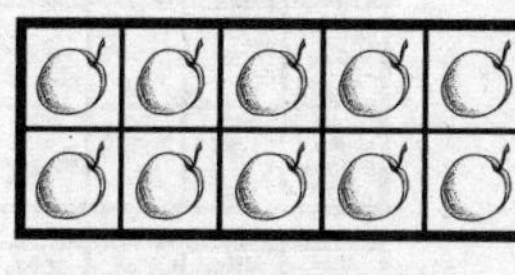

DIRECTIONS 1. Start with 14. Count forward. Write the numbers in order. **2.** Mark under all sets of numbers that are in counting order. **3.** What number does each set show? Write the numbers. Then write the numbers in counting order.

GO ON

Name ______________________

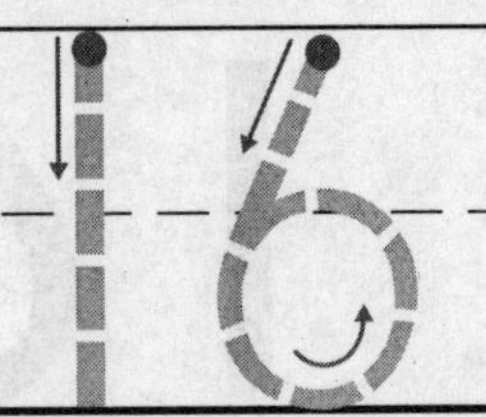

20
twenty

13 14 15 ○

11 15 12 ○

16 17 18 ○

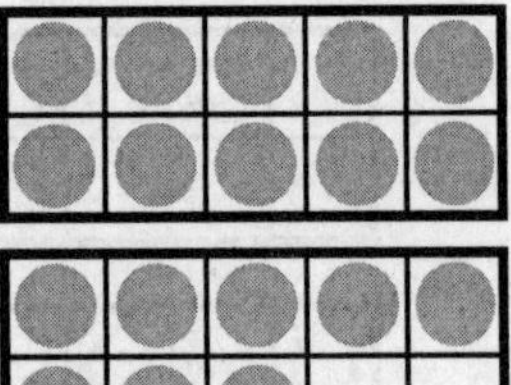

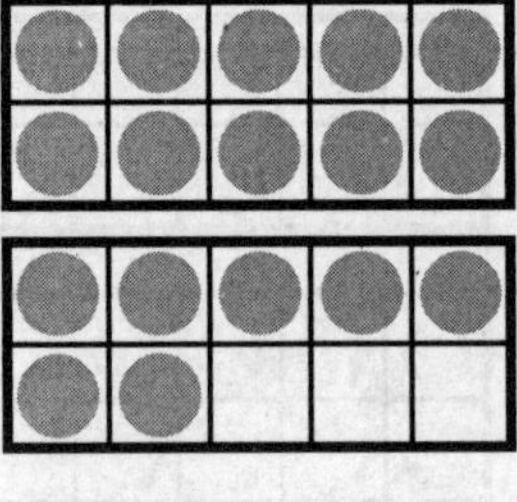

DIRECTIONS **4.** Start with 16. Count forward. Write the numbers in order. **5.** Mark under all sets of numbers that are in counting order. **6.** What number does each set of counters show? Write the numbers. Then write the numbers in counting order.

Name ______________________________

Practice Test
K.CC.A.3
Know number names and the count sequence.

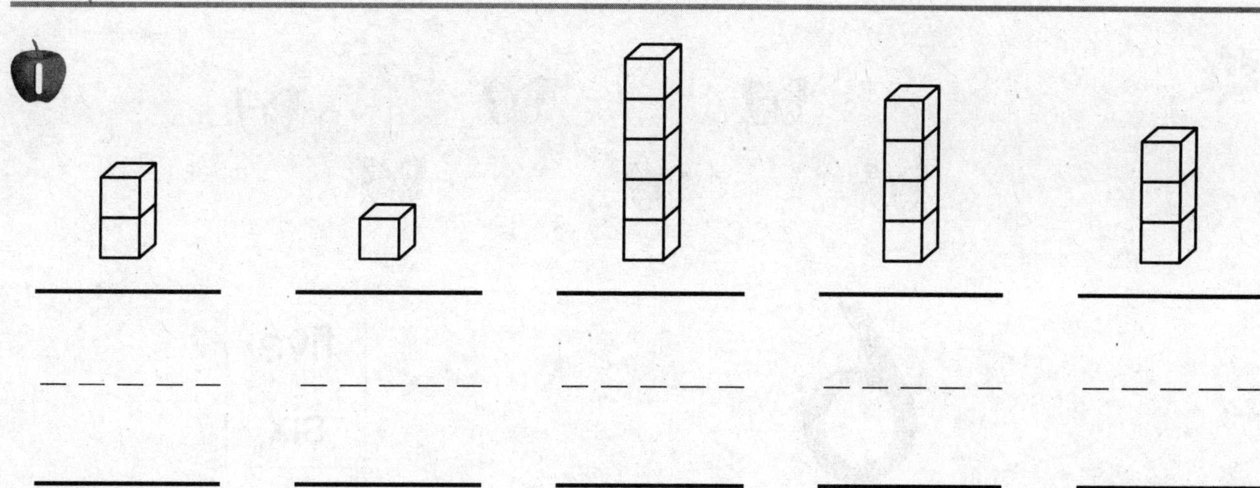

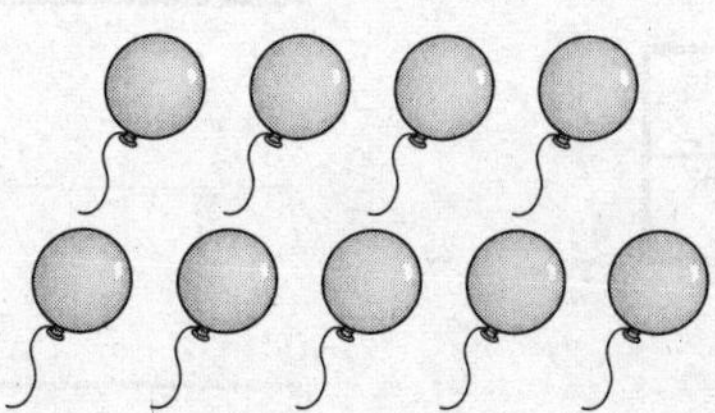

DIRECTIONS **1.** Count the cubes in each tower. Write the number. **2.** Four children each have one doll. Draw counters to show the dolls. Write the number. **3.** How many balloons are shown? Write the number.

GO ON

Name ____________________

6

five
six

5

____

hats

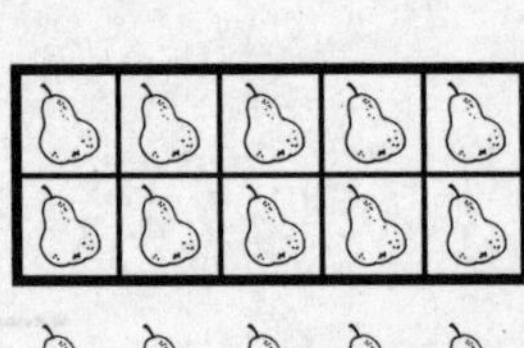
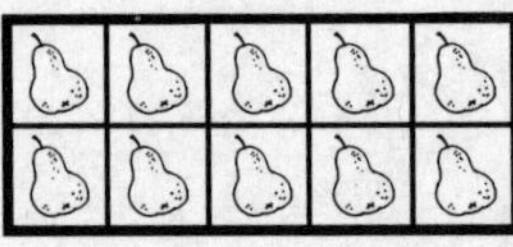

____

DIRECTIONS **4.** What is another way to write 6? Circle the word. **5.** There are 6 green hats and 7 blue hats. Draw the hats. Circle a group of 10. How many hats are there in all? **6.** Harry has 20 pears. Circle how many pears he has. Write the number of pears.

Name ____________________

Practice Test
K.CC.B.4a
Count to tell the number of objects.

- ○ 1
- ○ 2
- ○ two

- ○ 5
- ○ five
- ○ 4

 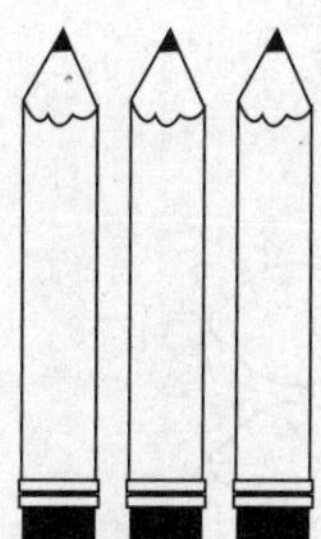 

DIRECTIONS 1–2. Choose all the answers that tell how many. **3.** Count how many. Write the number. **4.** Circle all the sets that show 3.

GO ON

Name ____________________

5

○ 1
○ 2
○ one

6

○ four
○ five
○ 5

7

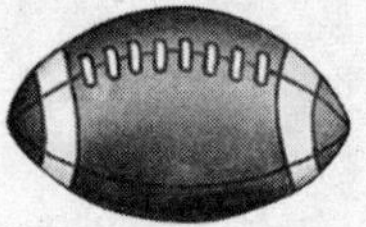
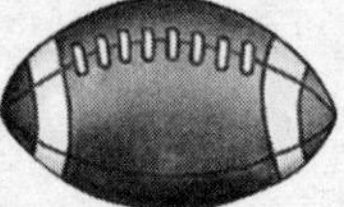
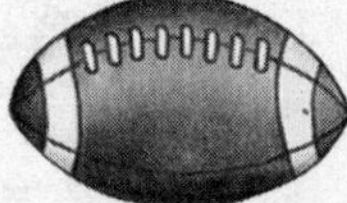

8

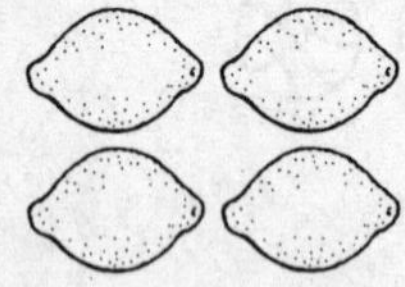

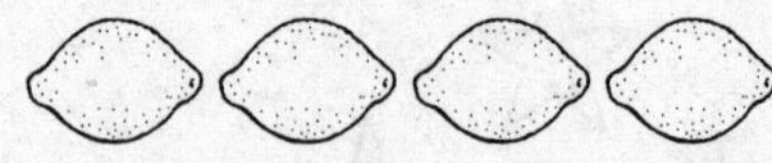

DIRECTIONS 5–6. Choose all the answers that tell how many. **7.** Count how many. Write the number. **8.** Circle all sets that show 4.

Name ______________________________

Practice Test
K.CC.B.4b
Count to tell the number of objects.

______ and ______

______ and ______

______

DIRECTIONS **1.** Show 2 ways to make 5. Color some boxes red. Color some boxes yellow. Write the numbers. **2.** Count how many. Write the number.

Name ______________________

3

_____ _____ _____ _____ _____

4

_____ and _____

_____ and _____

DIRECTIONS **3.** Count the cubes in each tower. Write the number. **4.** Show 2 ways to make 5. Color some counters red. Color some yellow. Write the numbers.

Name ______________________________

Practice Test
K.CC.B.4c
Count to tell the number of objects.

1

2

2 3 4 ○

4 3 5 ○

3 4 5 ○

3

DIRECTIONS 1. Write the numbers 1 to 5 in counting order. 2. Mark under all sets of numbers that are in counting order. 3. Write the number that comes after 4 in counting order. Draw counters to show the number.

GO ON

Name ____________________

4

5

4 2 1 ○ 3 4 5 ○ 1 2 3 ○

6

DIRECTIONS **4.** Write the numbers 6 to 10 in counting order. **5.** Mark under all sets of numbers that are in counting order. **6.** Write the number that comes after 3 in counting order. Draw counters to show the number.

Name ______________________________

Practice Test
K.CC.B.5
Count to tell the number of objects.

○ ○ ○

9 10 8

DIRECTIONS 1. Circle all the sets that show 6. **2.** Mark under all the sets that have 10 items. **3.** Match sets to the numbers that show how many fish.

GO ON

Name ____________________

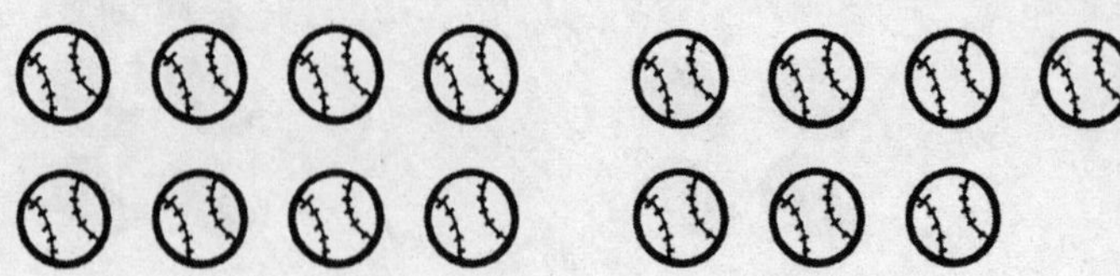

5

9

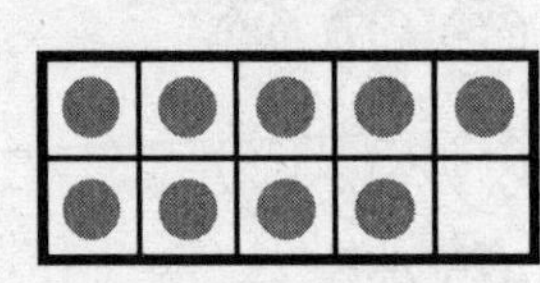

4 and | 4 / 5 | more

6

18 17 20

DIRECTIONS 4. Circle all the sets that show 7. 5. The ten frame shows 4 counters on the bottom and some on the top. Four and how many more make 9? Choose the number. 6. Match the ten frames to the numbers that tell how many cubes.

Name ______________________________

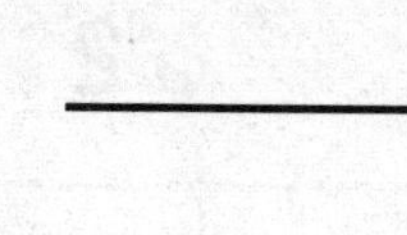

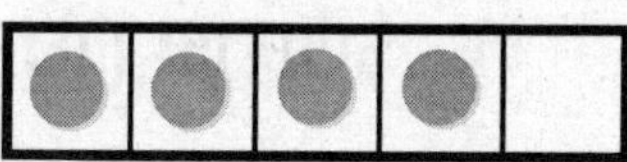

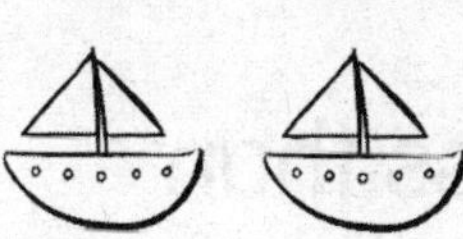

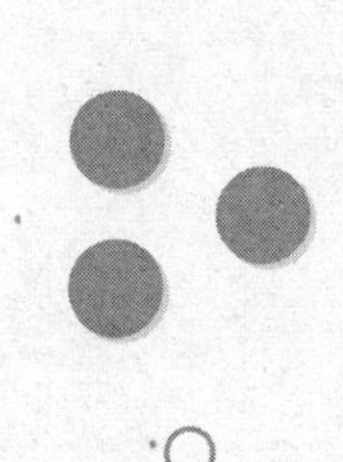

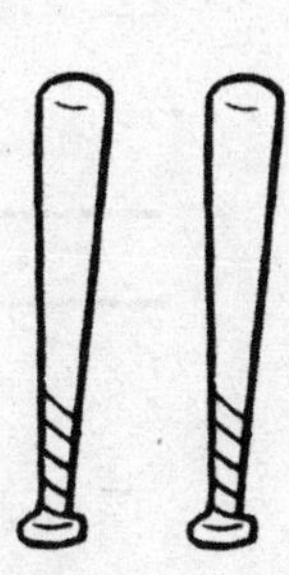

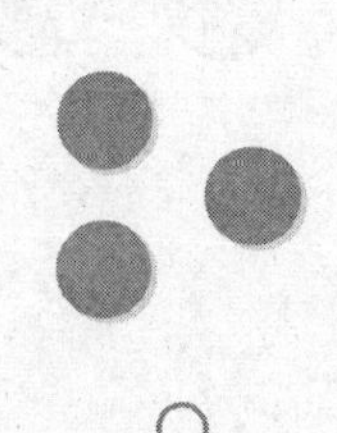

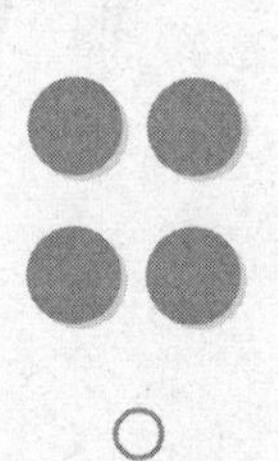

DIRECTIONS **1.** How many counters are there in each row? Write the numbers. Compare the sets by matching. Circle the number that is greater. **2.** Mark under all the sets that have the same number of counters as the number of boats. **3.** Mark under all the sets that have a number of counters greater than the number of bats.

GO ON

Name ______________________________

4

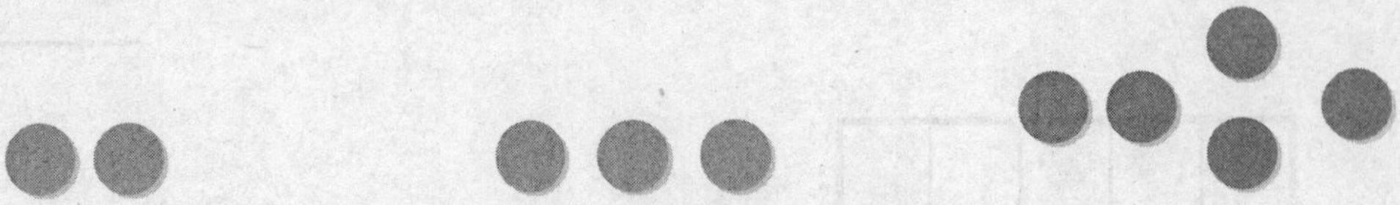

○ 2 ○ 3 ○ 5

5

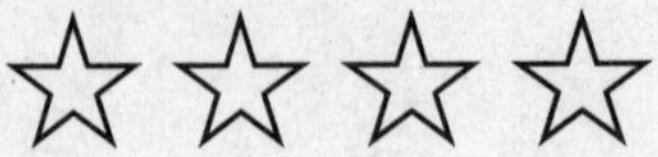

same number

greater than

less than

6

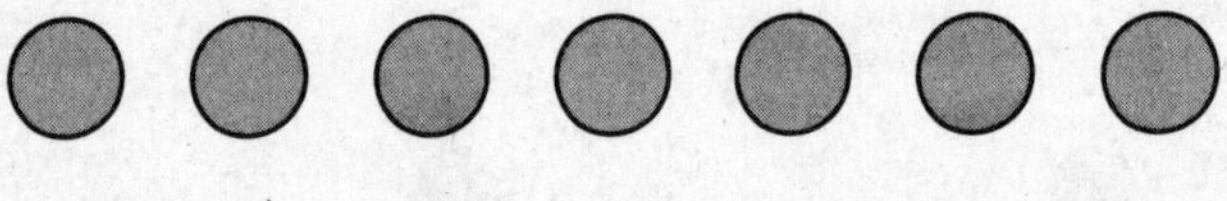

DIRECTIONS **4.** Mark all the numbers less than 5. **5.** Compare the number of counters in each set to the number of stars. Draw lines from the sets of counters to the words that show *same number, greater than,* or *less than.* **6.** Write how many counters are in the set. Use matching lines to draw a set of counters less than the number of counters shown. Circle the number that is less.

Name ________________________________

Practice Test
K.CC.C.7
Compare numbers.

○ ○ ○

6 

7 9

DIRECTIONS **1.** Choose all the cube towers that have a number of cubes greater than 6. **2–3.** Look at the numbers. Think about the counting order as you compare the numbers. Circle the greater number.

GO ON

Name ______________________________

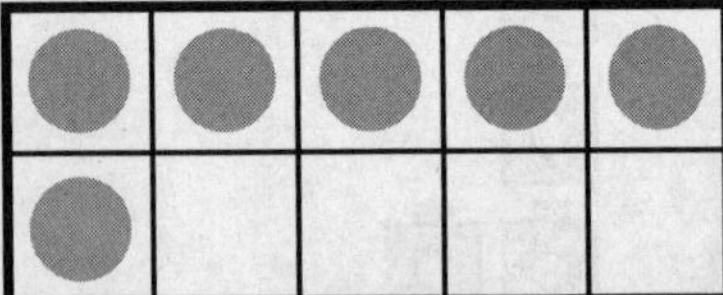

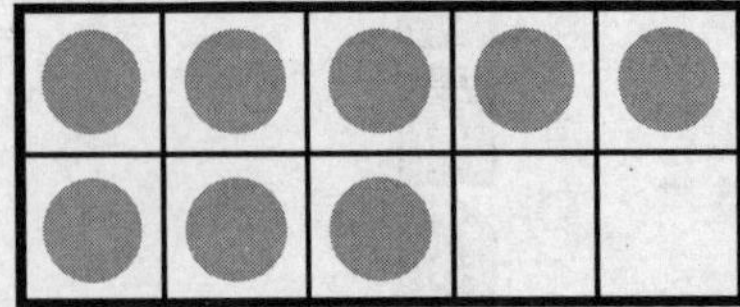

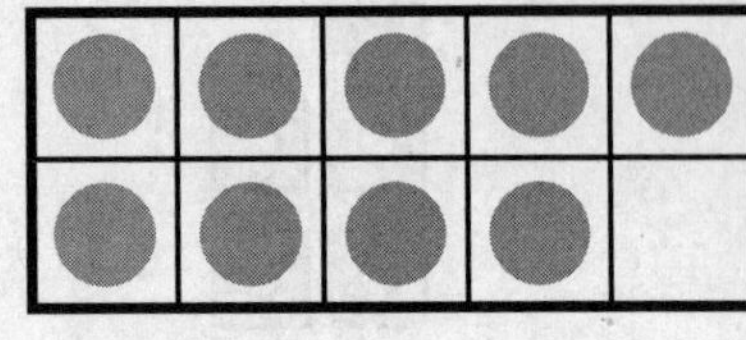

○ ○ ○

5 3

6

8 9

DIRECTIONS **4.** Choose all the ten frames that have a number of counters greater than 6. **5–6.** Look at the numbers. Think about the counting order as you compare the numbers. Circle the number that is less.

Name ________________________________

Practice Test

K.OA.A.1
Understand addition, and understand subtraction.

______ and ______

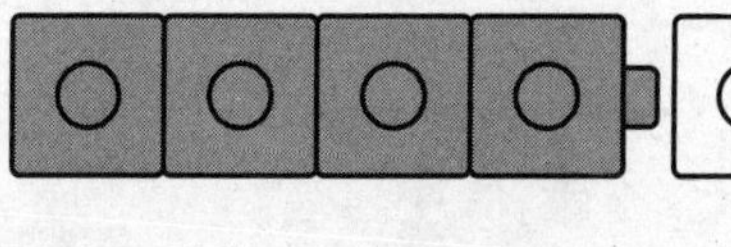

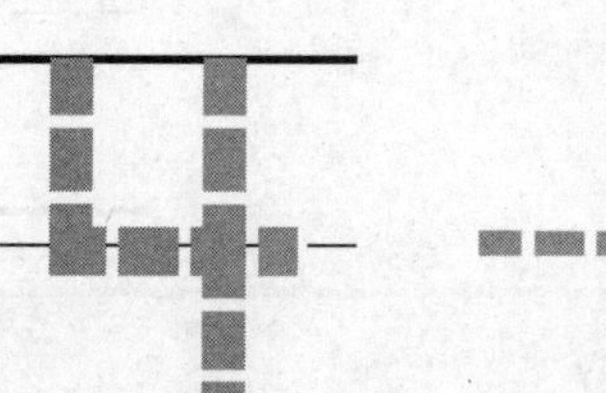

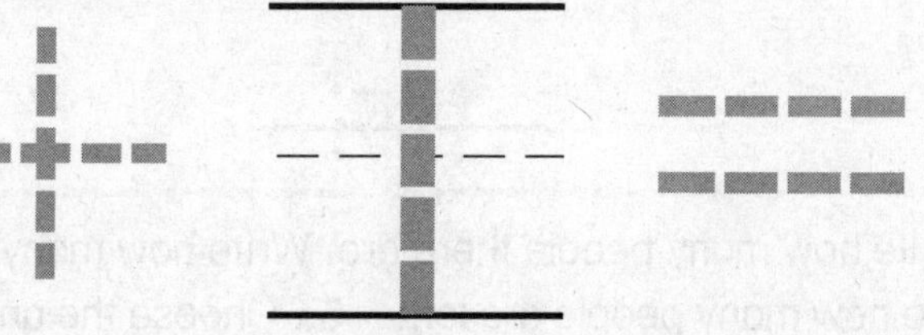

DIRECTIONS I. How many children are facing right? How many children are being added to the group? Write the numbers. **2.** Draw a picture that shows 7 + 1 squares. Write how many squares. **3.** How many of each color cube is being added? Trace the numbers and symbols. Write the number that shows how many cubes in all.

Name ______________________

4

______ take away ______ ______

5

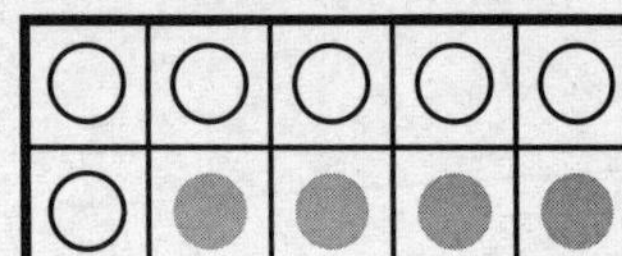

- ○ 10 − 4
- ○ 6 − 4
- ○ 10 − 1

6

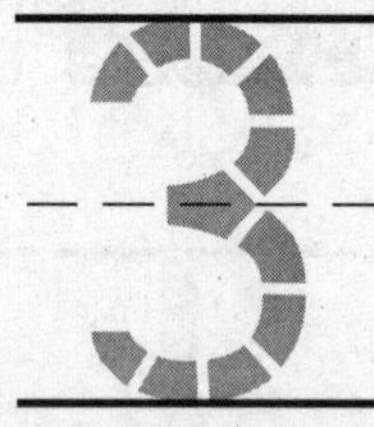

5 − 3 = ______

DIRECTIONS 4. Write how many people there are. Write how many people are leaving. Write how many people are left. 5. Choose the answer that shows how many counters are white. 6. Model a five-cube train. Three cubes are gray and the rest are white. Take apart the cube train to show how many are white. Draw the cube trains. Trace and write to complete the subtraction sentence.

Name ________________________________

Practice Test
K.OA.A.2
Understand addition, and understand subtraction.

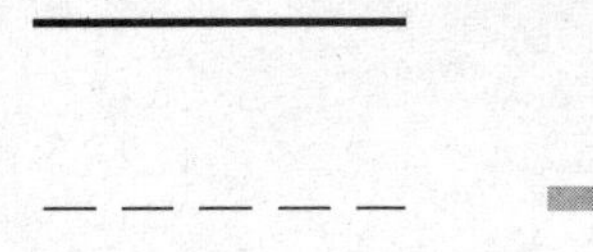
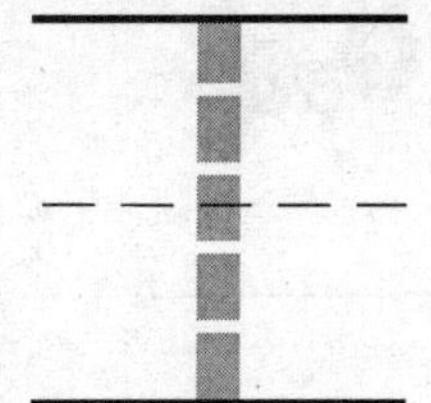

___ − 1 = 2

8 − 1 = 6 / 7 / 8

10 = 6 + 4 ○ 8 = 4 + 4 ○ 4 + 6 = 10 ○

DIRECTIONS 1. There are some cats. One is taken from the set. How many cats were there to start? Write and trace to complete the subtraction sentence. **2.** There are 8 circles. Draw a set of circles that shows 8 − 1. Circle the number that completes the number sentence. **3.** Mark under all the number sentences that match the cubes.

GO ON

Name ______________________

4

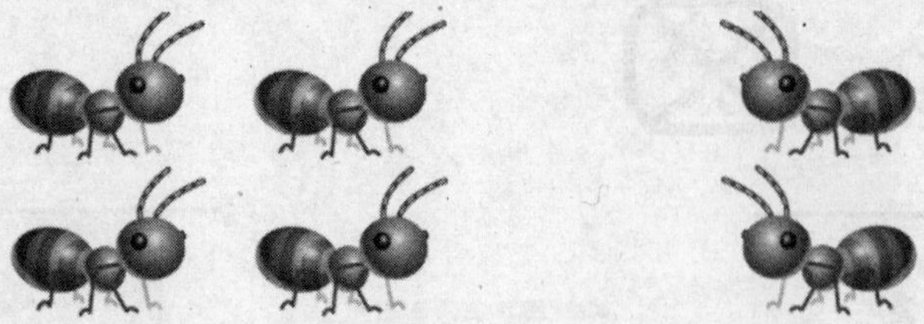

4 + ___ = 6

5

___ − ___ = 0

6

___ − 3 = 1

DIRECTIONS 4. Write the numbers and trace the symbols to complete the addition sentence. 5. Arthur had some grapes. He ate some grapes. Now there are zero grapes left. Draw to show how many grapes there could have been to start. Cross out grapes to show how many were eaten. Complete the subtraction sentence. 6. There are some butterflies. Three of the butterflies are taken from the set. Draw more butterflies to show how many butterflies there were to start. Write the number to complete the subtraction sentence.

Name ______________________________

Practice Test
K.OA.A.3
Understand addition, and understand subtraction.

1 + 4	4 + 4	5 + 1
○	○	○

DIRECTIONS 1. Ryan has 2 big marbles. Dani has some small marbles. Together they have 5 marbles. Draw to show how many small marbles Dani has. Complete the number pair. **2.** Choose the number pair that makes a number greater than 6. **3.** Complete the addition sentence to show a number pair for 10.

Name ______________________

9 =  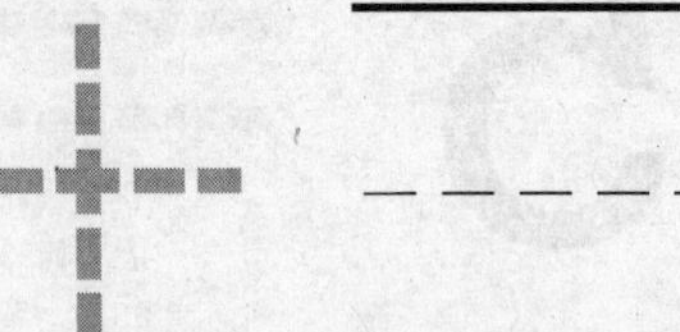

5 = ____ + ____

5 + 2 ○   1 + 3 ○   2 + 4 ○

DIRECTIONS **4.** Paul has 9 tokens. Each token is either red or blue. How many red and blue tokens could he have? Color the tokens to show the number of red and blue tokens. Write the numbers to complete the addition sentence. **5.** Nora has 1 green crayon. Gary has some red crayons. Together they have 5 crayons. Draw to show how many red crayons Gary has. Complete the number pair. **6.** Choose the number pair that makes a number greater than 6.

Name ______________________________

Practice Test
K.OA.A.4
Understand addition, and understand subtraction.

1

cubes

2

3

5 + = 10

DIRECTIONS **1.** Write how many gray cubes. Write how many white cubes. Write how many cubes in all. **2.** How many counters are there? Write the number. How many more counters do you need to make 10? Write the number. **3.** Look at the cube train. How many gray cubes do you see? How many more cubes are added to make 10? Draw the cubes. Write and trace to show this in an addition sentence.

GO ON

Name ______________________

cubes

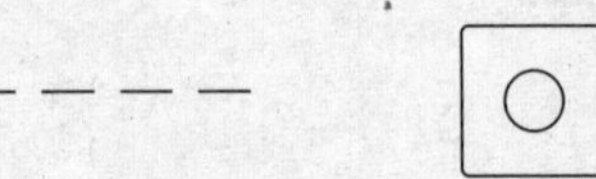

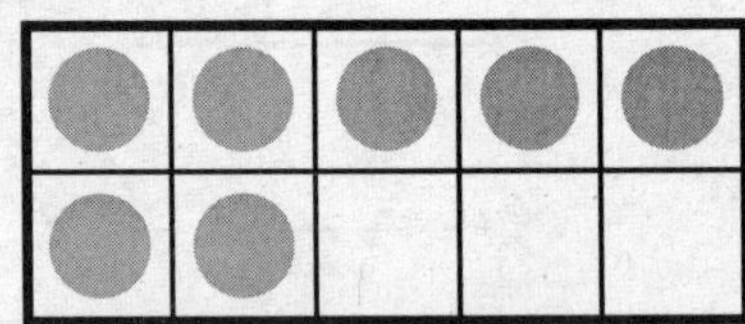
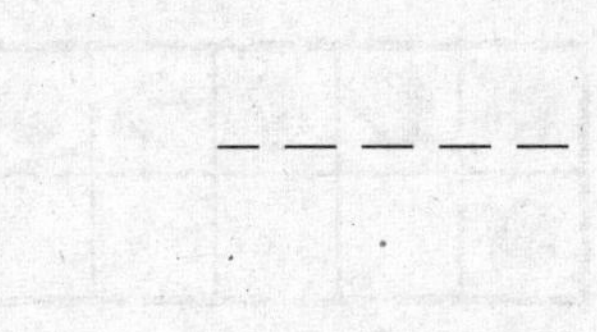

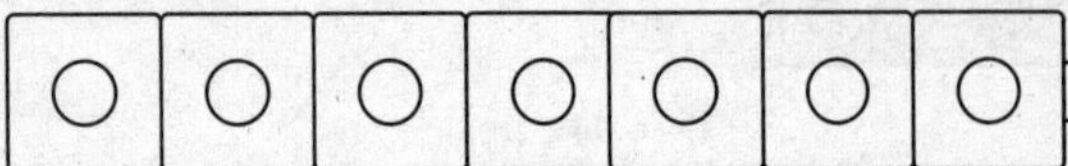

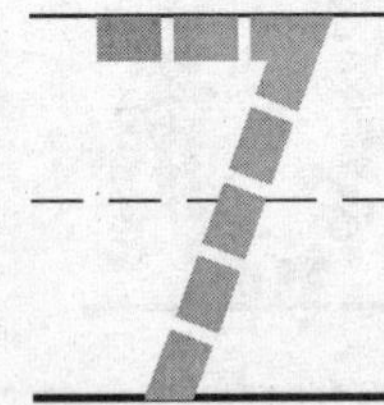
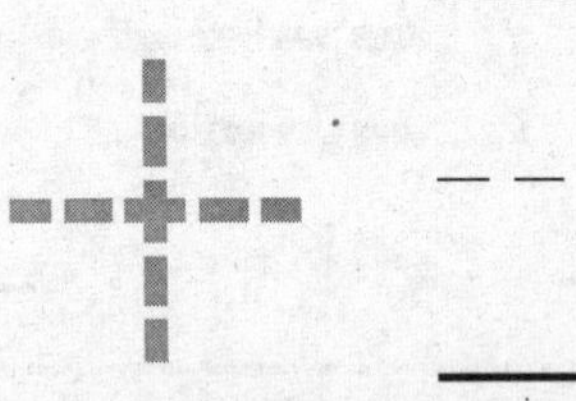

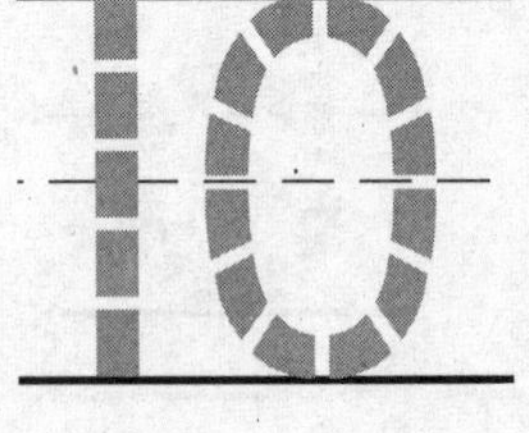

DIRECTIONS **4.** Write how many gray cubes. Write how many white cubes. Write how many cubes in all. **5.** How many counters are there? Write the number. How many more counters do you need to make 10? Write the number. **6.** Look at the cube train. How many cubes do you see? How many more cubes do you need to add to make 10? Draw the cubes. Write and trace to show this as an addition sentence.

Name ______________________________

Practice Test
K.OA.A.5
Understand addition, and understand subtraction.

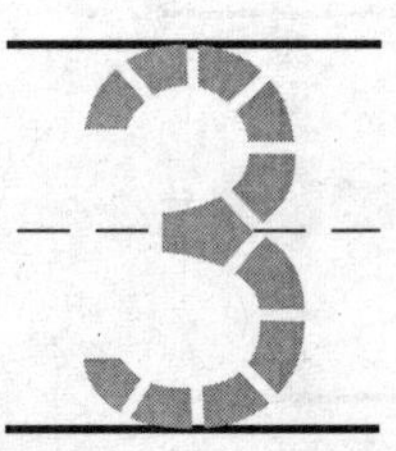

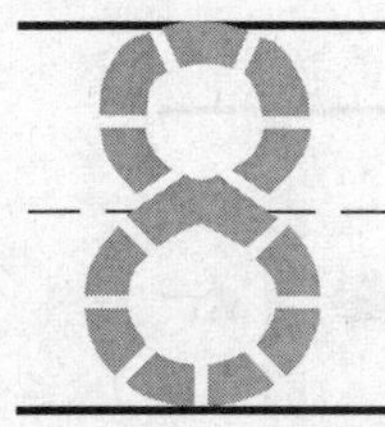

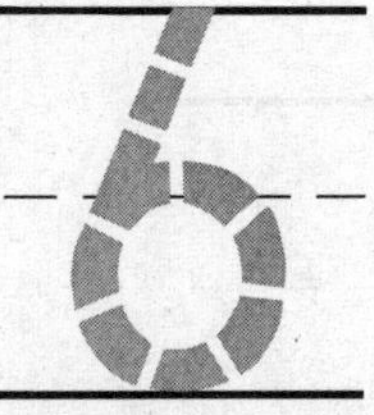
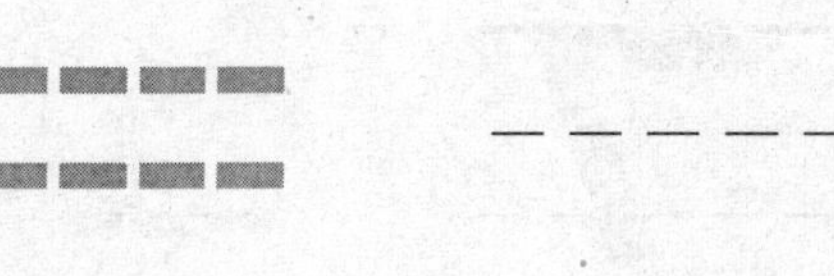

DIRECTIONS **1.** Jeff has 2 red cubes. He has 3 yellow cubes. How many cubes does he have? Draw the cubes. Trace the numbers and symbols. Write how many in all. **2.** There are 5 horses. Some horses are taken from the set. Trace and write to complete the subtraction sentence. **3.** Model an eight-cube train. Six cubes are gray and the rest are white. Take apart the cube train to show how many are white. Draw the cube trains. Complete the subtraction sentence.

GO ON

Name ____________________

4

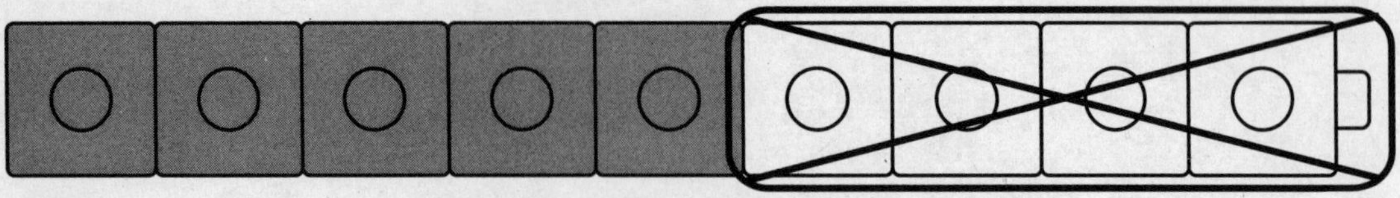

9 – 4 = ___

5

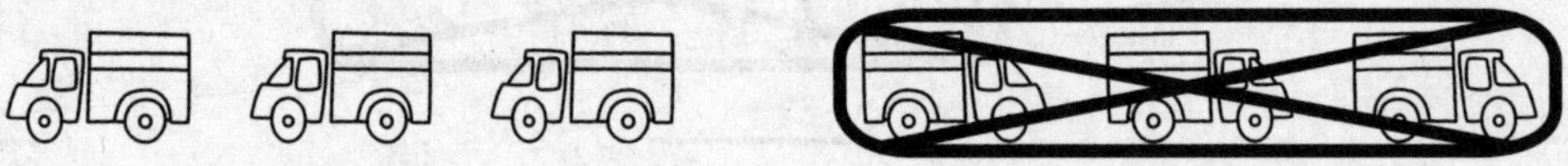

6 – ___ = ___

6

7 – ___ = 4

DIRECTIONS **4–5.** Complete the subtraction sentence to match the picture. **6.** Mabel started a game with 7 baseballs. Some of the baseballs were lost. Now Mabel has 4 baseballs. How many baseballs were lost? Draw to solve the problem. Complete the subtraction sentence.

STOP

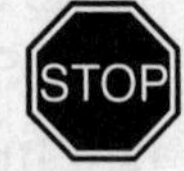

Name ________________________________

Practice Test
K.NBT.A.1
Work with numbers 11-19 to gain foundations for place value.

1

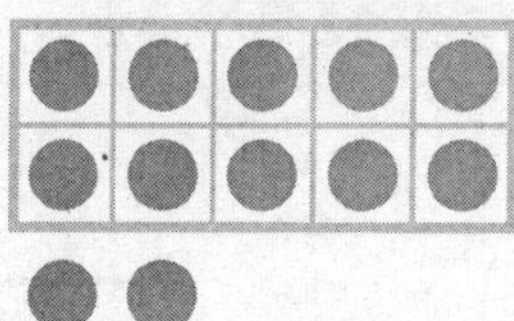

2

10 + 1

3

○ 15
○ 14
○ 10 + 5

4

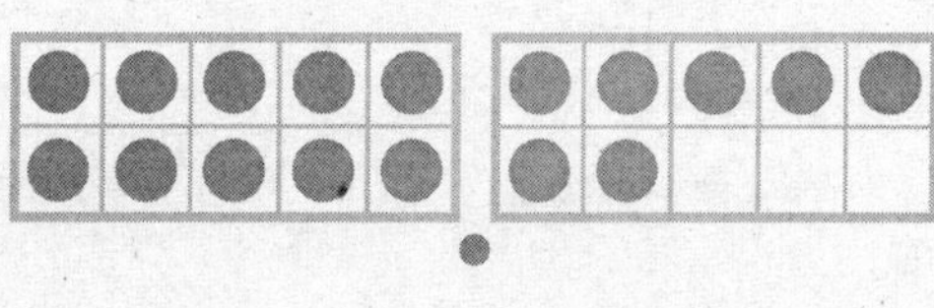

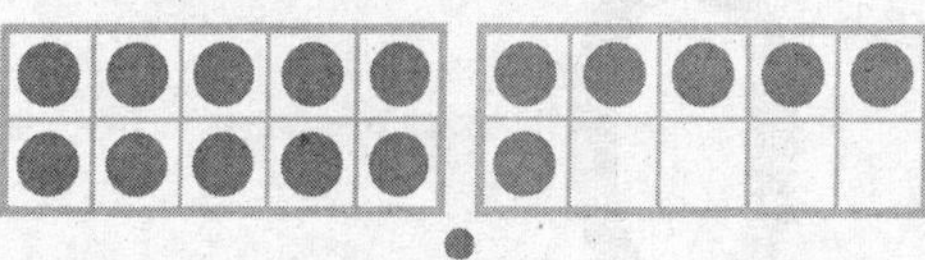

16 17

DIRECTIONS 1. How many objects are there? Write the number. **2.** Choose all the ways that show 11. **3.** Choose the way to write the number of bears in the set. **4.** Draw lines to match the ten frames to the numbers they show.

Name ____________________

5

☆☆☆☆☆
☆☆☆☆☆
☆☆☆☆☆

______

______ + ______ = ______

6

______ + ______ = ______

7

10 + ______ = ______

DIRECTIONS **5.** Count how many. Write the number. Complete the addition sentence. **6.** Look at the ten frames. Complete the addition sentence. **7.** Franklin has 8 blue buttons and 5 green buttons. Draw the buttons. Circle a group of 10 buttons. Count the remaining buttons starting from 10. Complete the addition sentence.

Name ______________________________

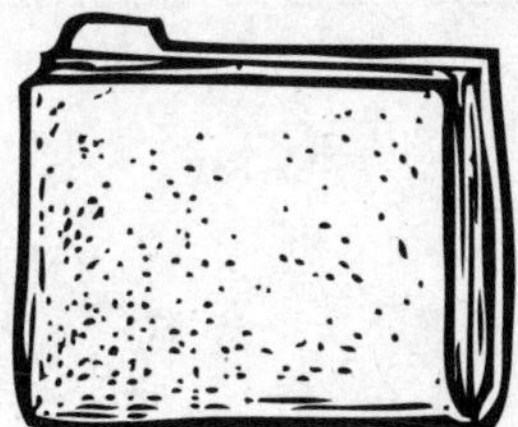

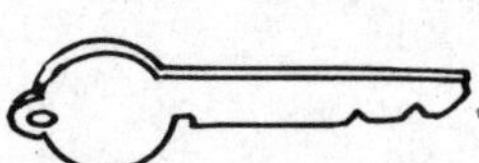

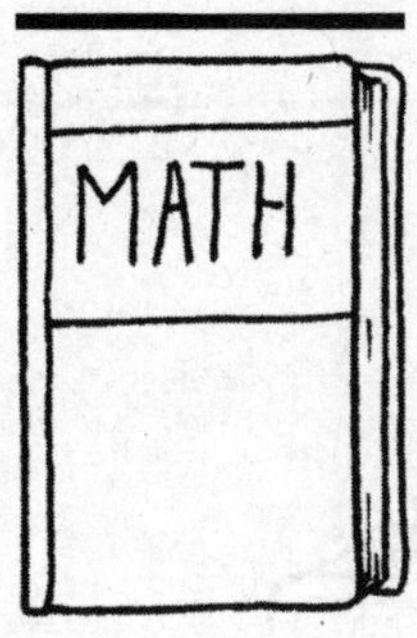

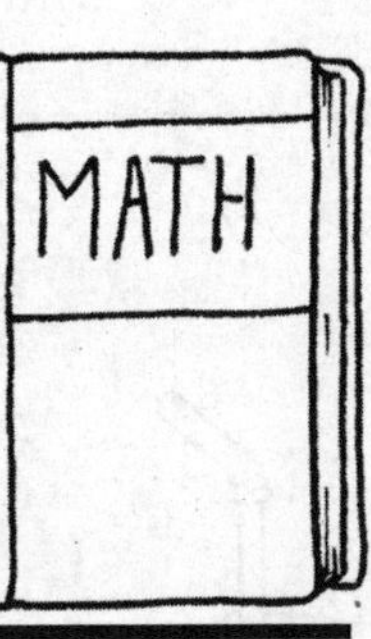

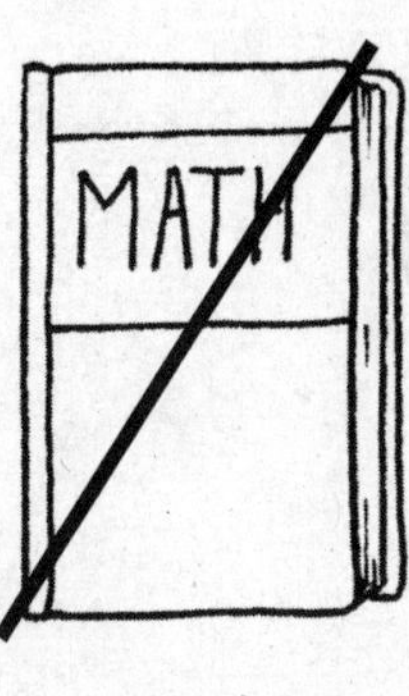

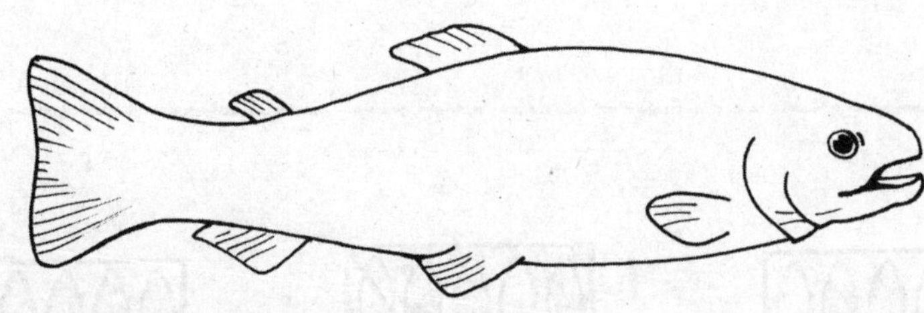

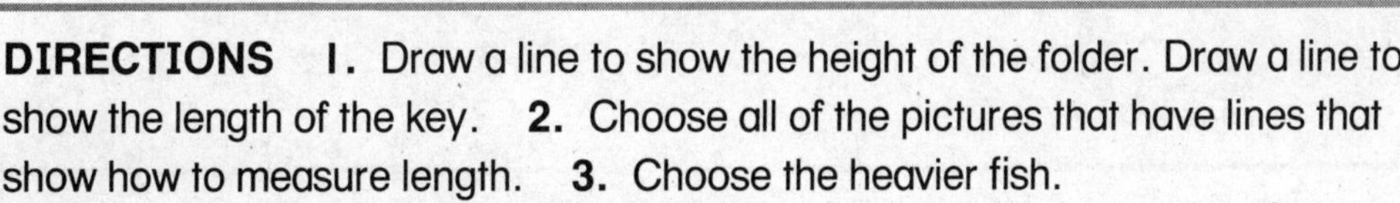

DIRECTIONS **1.** Draw a line to show the height of the folder. Draw a line to show the length of the key. **2.** Choose all of the pictures that have lines that show how to measure length. **3.** Choose the heavier fish.

GO ON

Name ______________________________

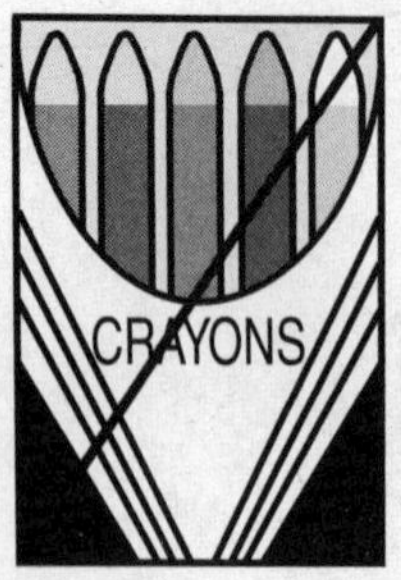

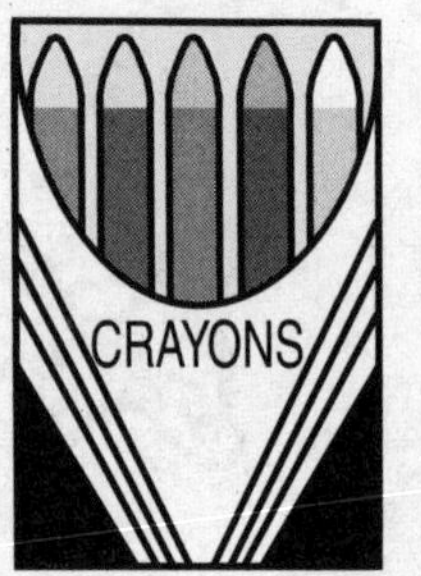

○ ○ ○ ○

DIRECTIONS 4. Draw a short flower and a tall flower. Circle the shorter one. 5. Draw a line to show the height of the juice box. Draw a line to show the length of the lunch box. 6. Choose all of the pictures that have lines that show how to measure height.

Name ______________________

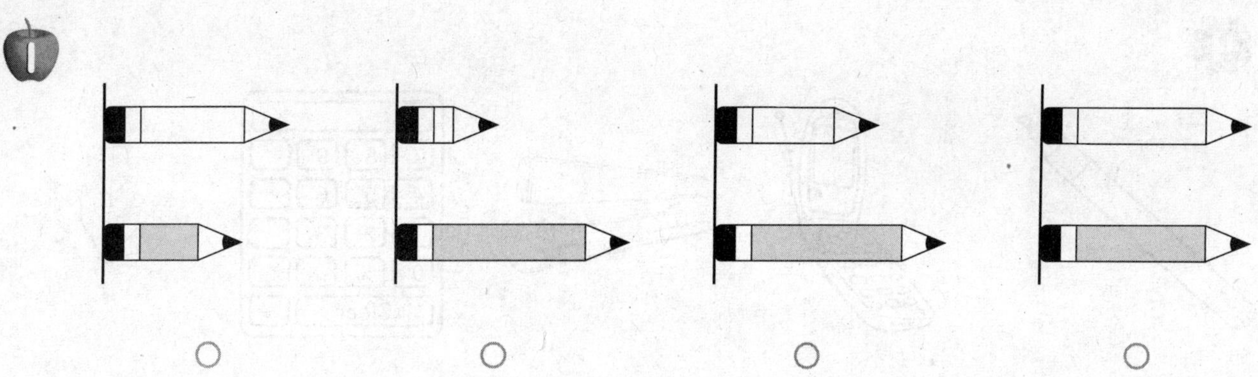

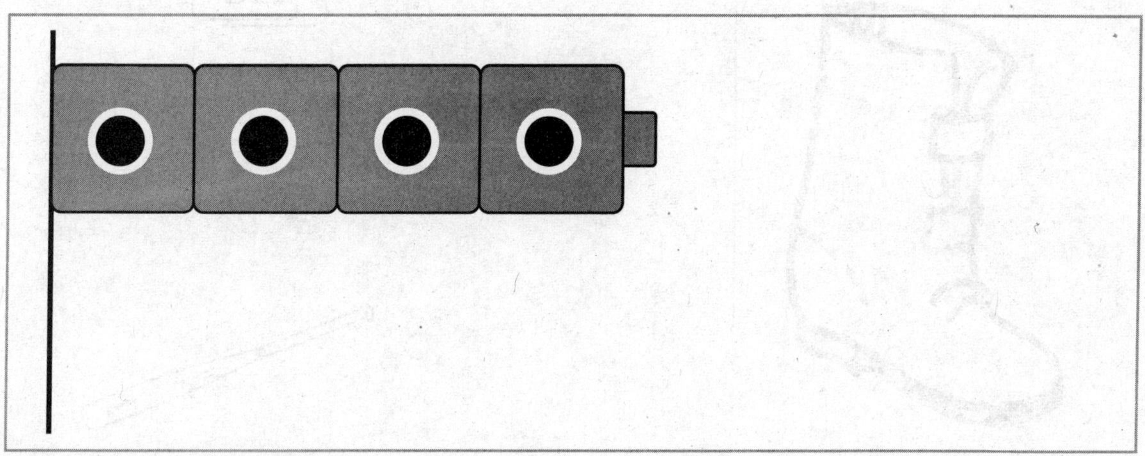

DIRECTIONS **1.** Choose all the sets that have a white pencil that is shorter than the gray pencil. **2.** Draw a cube train that is longer. **3.** Draw two pieces of string of different lengths. Draw a circle around the string that is shorter.

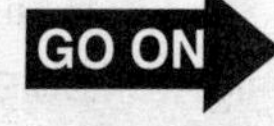

Name ____________________

4

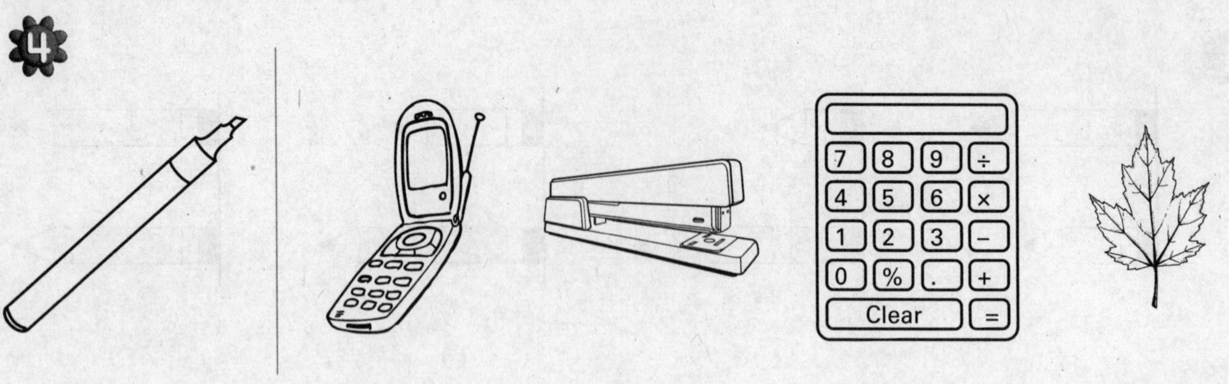

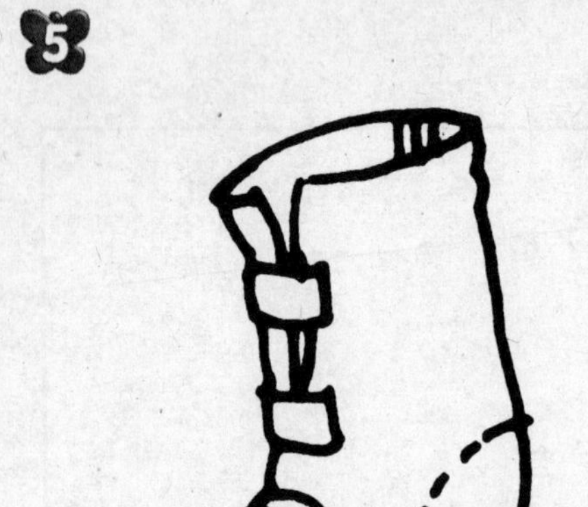

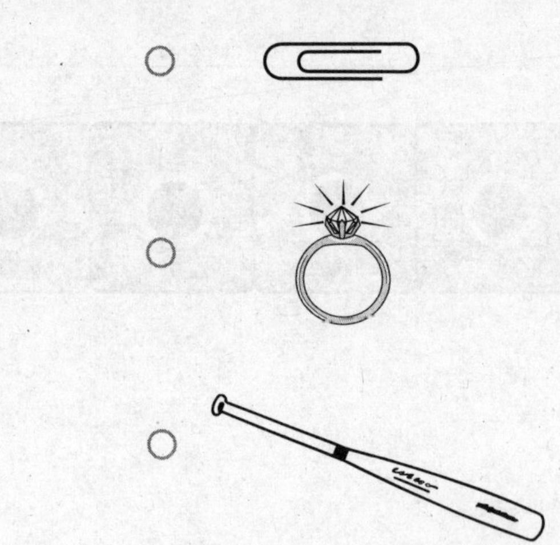

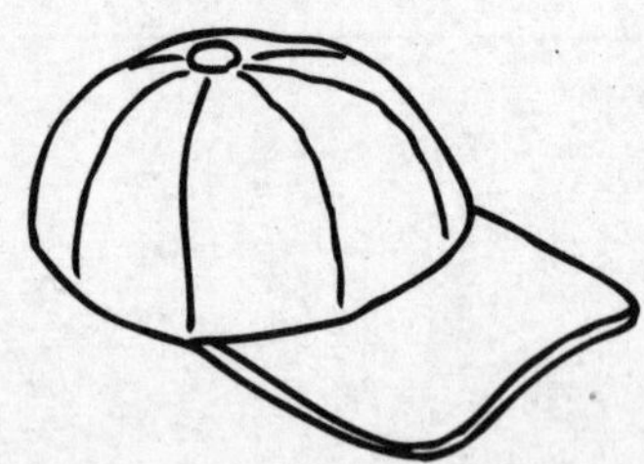

DIRECTIONS 4. Circle all the objects that are heavier than a marker. **5.** Choose all the objects that are lighter than the boot. **6.** Look at the objects. Circle the heavier object. Mark an X on the lighter object.

Name ______________________

Practice Test
K.MD.B.3
Classify objects and count the number of objects in each category.

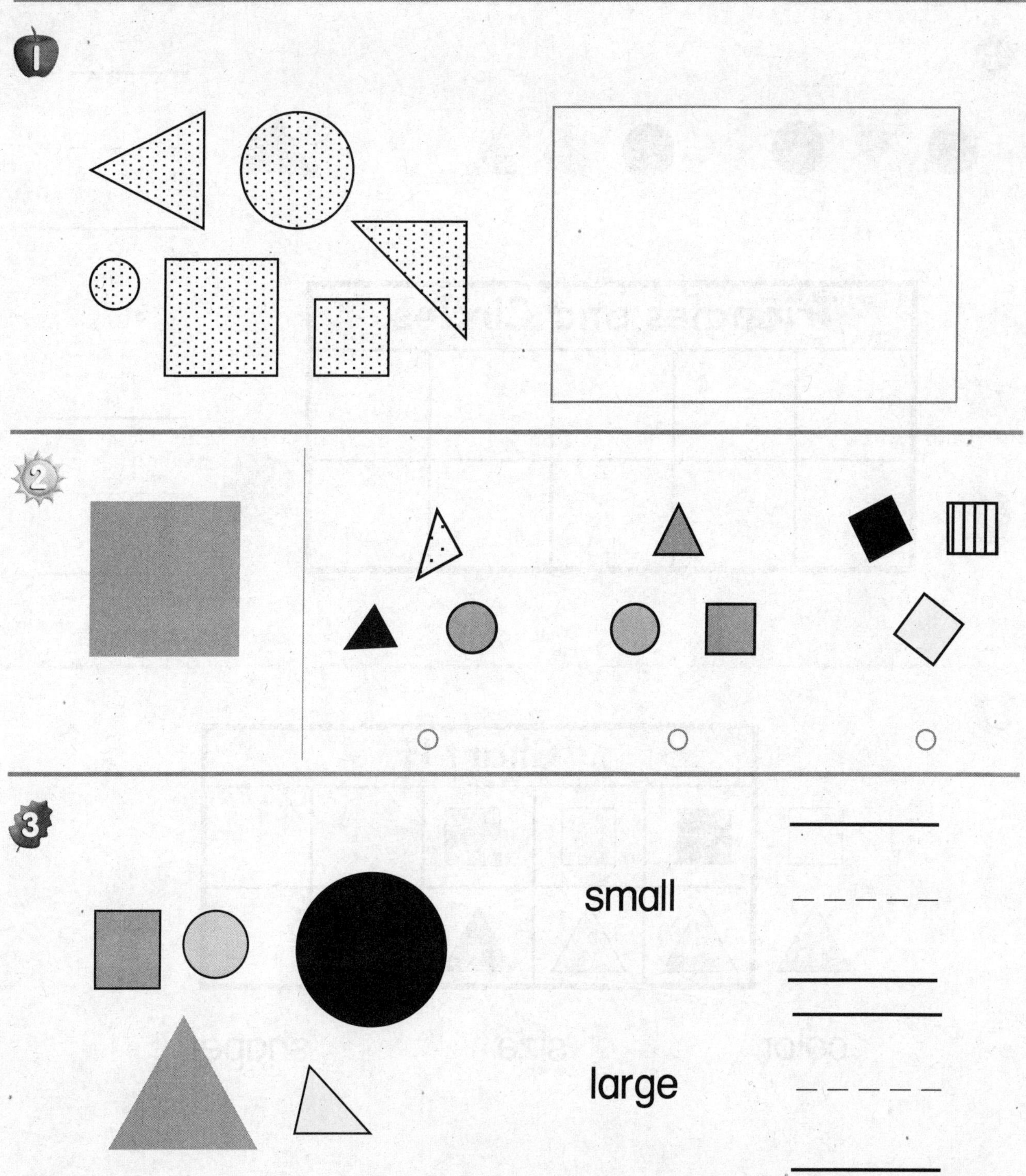

DIRECTIONS **1.** Draw and color a shape that belongs in this category. **2.** Look at the shape at the beginning of the row. Mark under all of the categories where the shape can belong. **3.** Draw a circle around each small shape. Write how many small shapes. Mark an X on each big shape. Write how many large shapes.

GO ON

Name ______________________

4

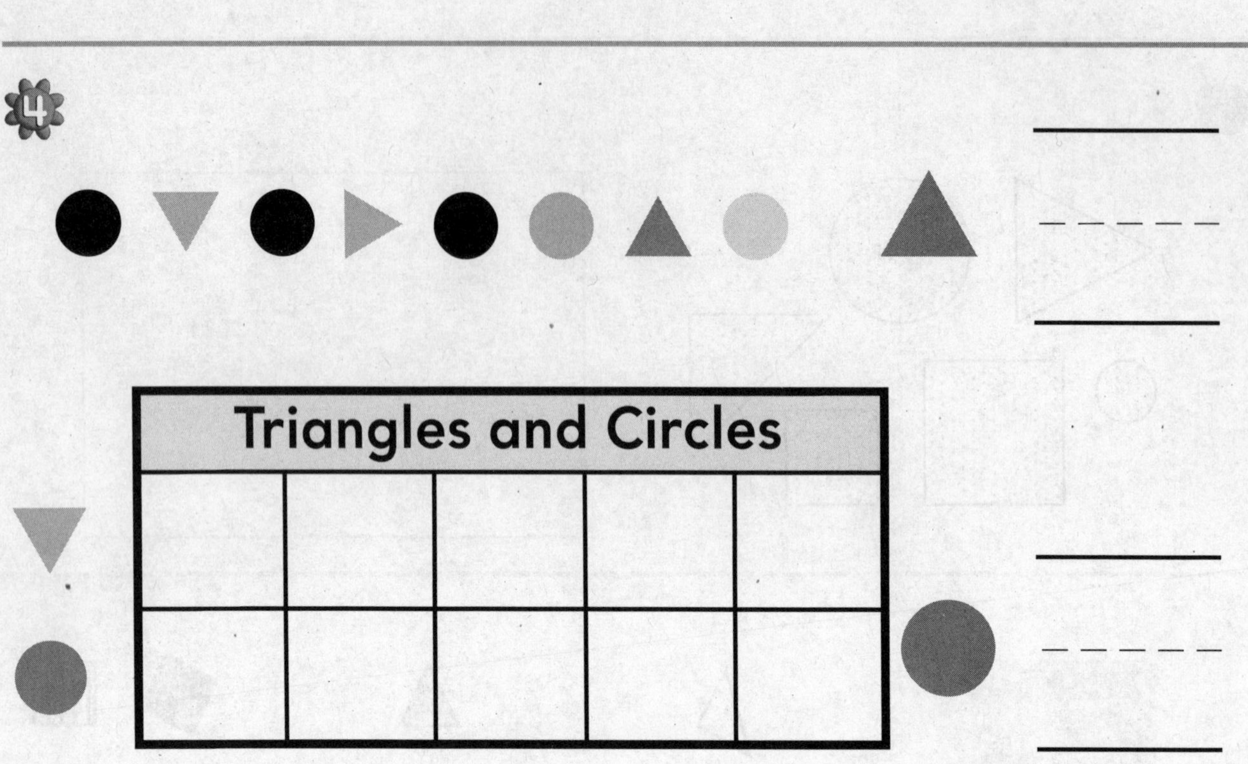

5

Chart H

color ○

size ○

shape ○

DIRECTIONS **4.** Sort and classify the shapes by category. Draw each shape on the graph. Write how many of each shape. **5.** Choose all the ways the chart is sorted.

Name ____________________

Practice Test
K.G.A.1
Identify and describe shapes.

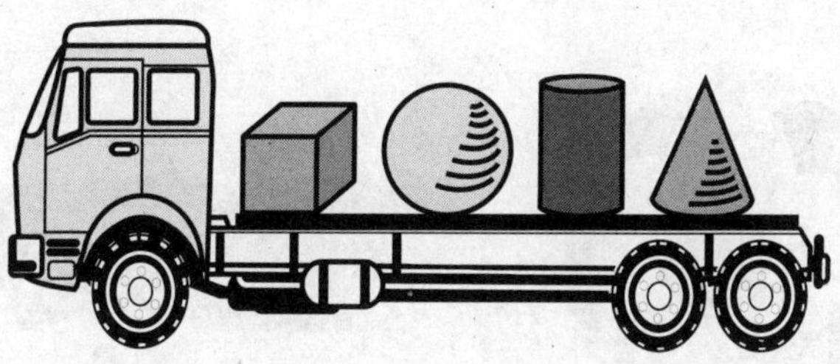

DIRECTIONS **1.** Mark an X on the shape that is next to the cylinder. **2.** Mark an X on the shape that is behind the sphere. **3.** Mark an X on the bead shaped like a cylinder that is next to the bead shaped like a cone.

GO ON

Name ______________________________

DIRECTIONS **4.** Mark an X on the object that is above the basketball net. **5.** Mark an X on the shape that is next to the cone. **6.** Mark an X on the object in front of the cube.

Name ______________________

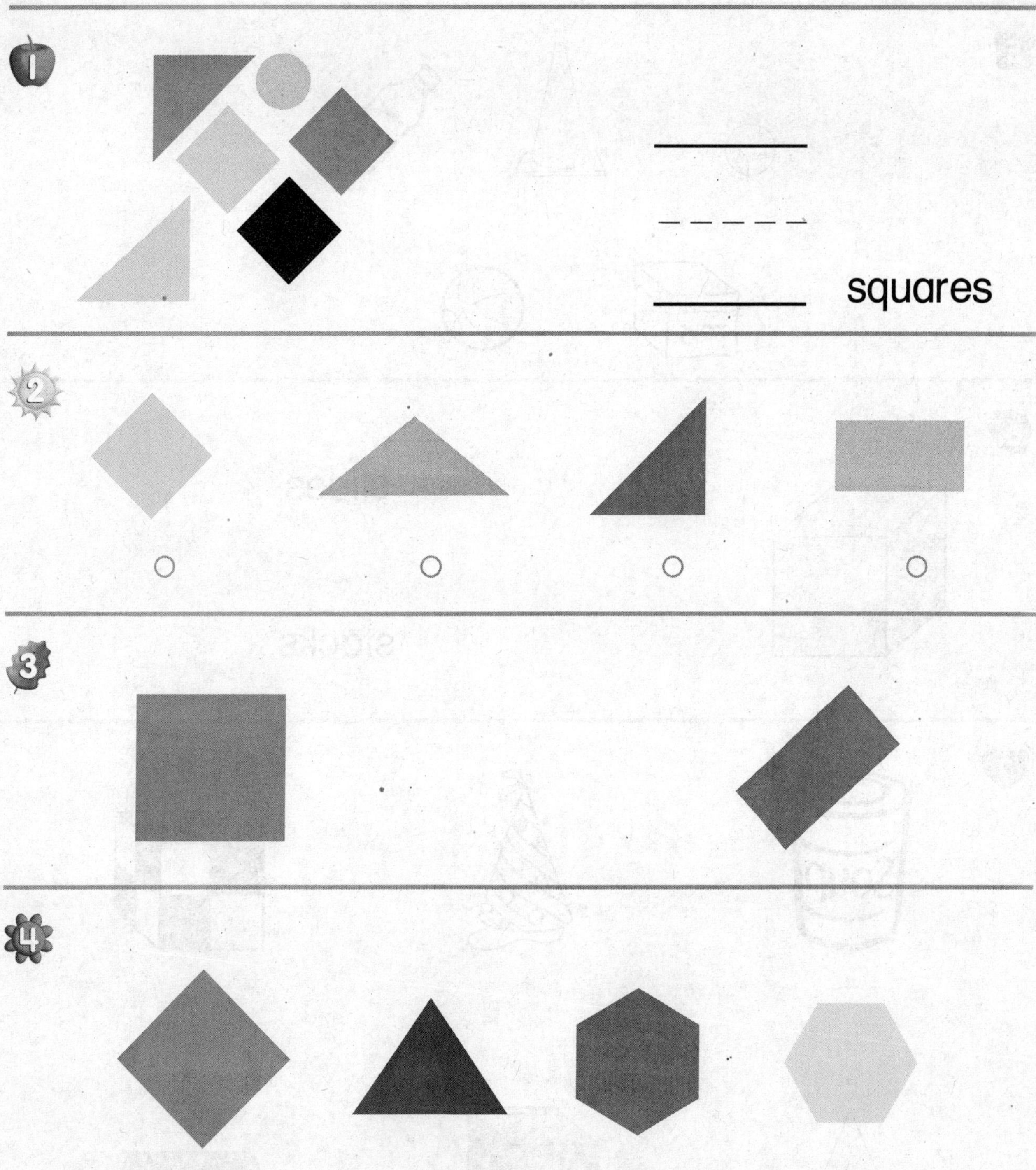

DIRECTIONS **1.** How many squares are in the picture? Write the number. **2.** Mark under all of the shapes that are triangles. **3.** Mark an X on the shape that is not a rectangle. **4.** Mark an X on all of the hexagons.

GO ON

Name ______________________________

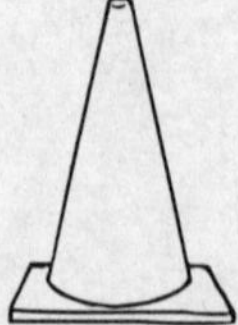

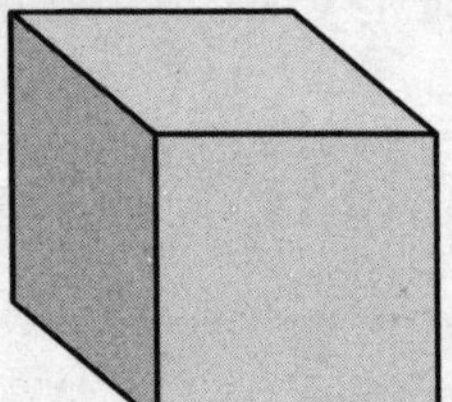

slides

stacks

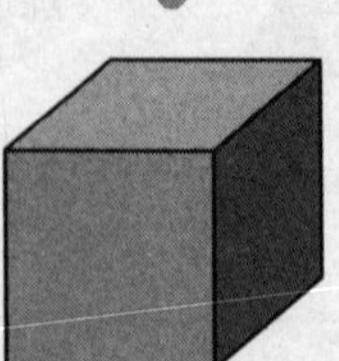

DIRECTIONS **5.** Which objects are shaped like a sphere? Mark an X on each of those objects. **6.** Circle the words that describe a cube. **7.** Draw lines to match the objects to their shapes.

Name ____________________

Practice Test

K.G.A.3

Identify and describe shapes.

1

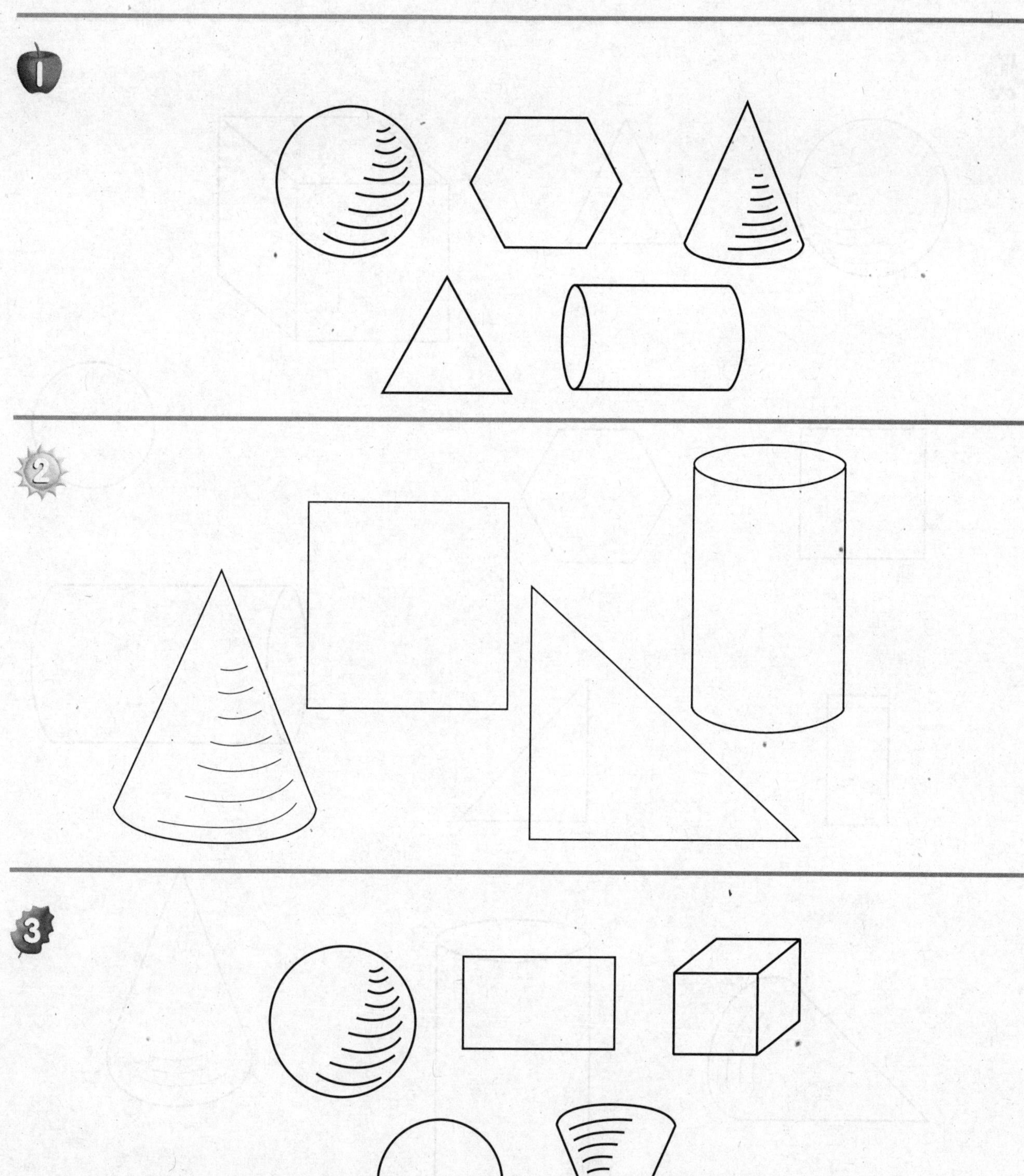

DIRECTIONS **1–2.** Color the solid shapes blue. Color the flat shapes red. Draw another flat shape that is different. **3.** Circle the solid shapes. Mark an "X" on the flat shapes.

GO ON

Name ______________________________

4

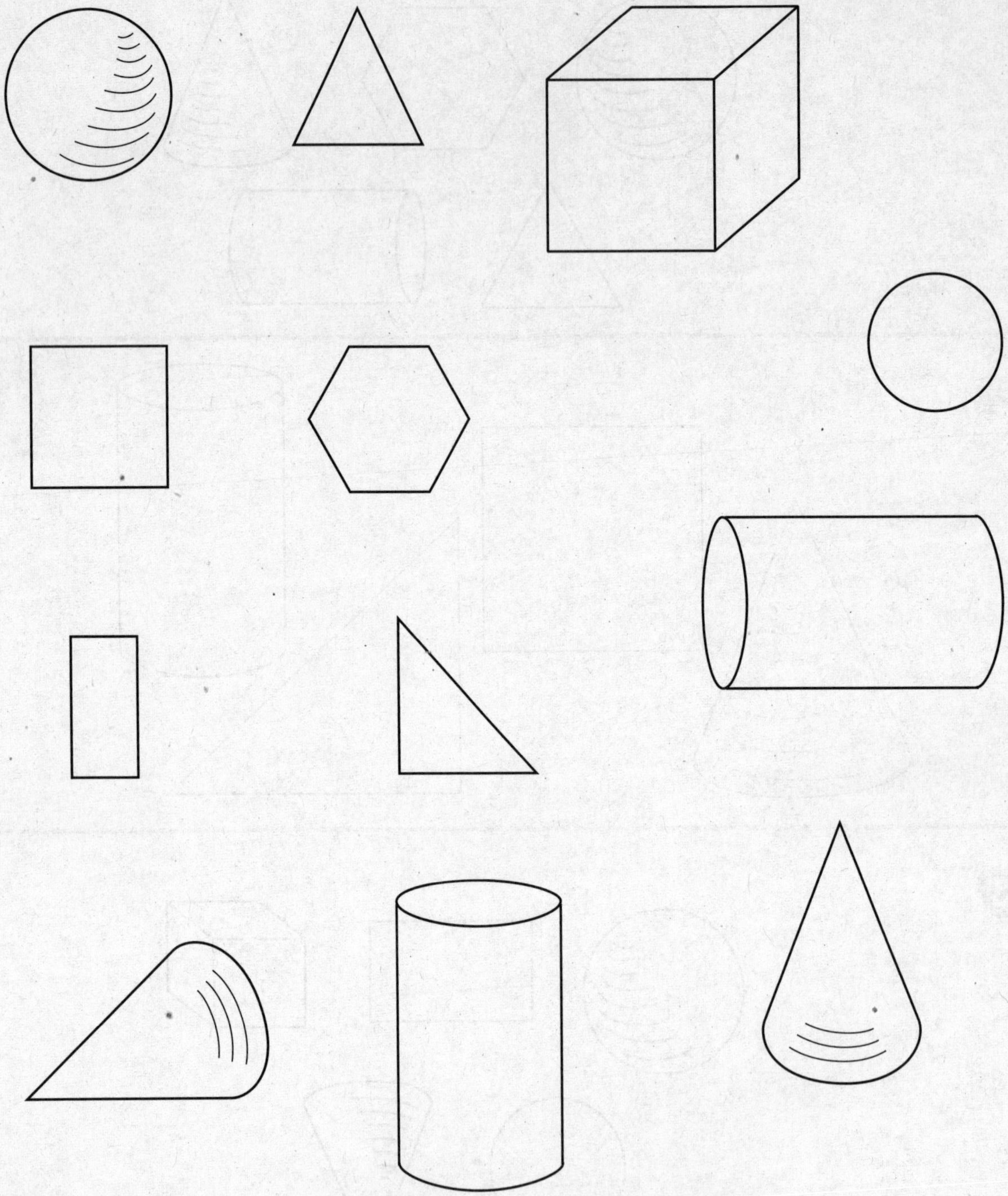

DIRECTIONS 4. Identify the two-dimensional or flat shapes. Use red to color the flat shapes. Identify the three-dimensional or solid shapes. Use blue to color the solid shapes.

STOP

Name ______________________________

Practice Test
K.G.B.4
Analyze, compare, create, and compose shapes.

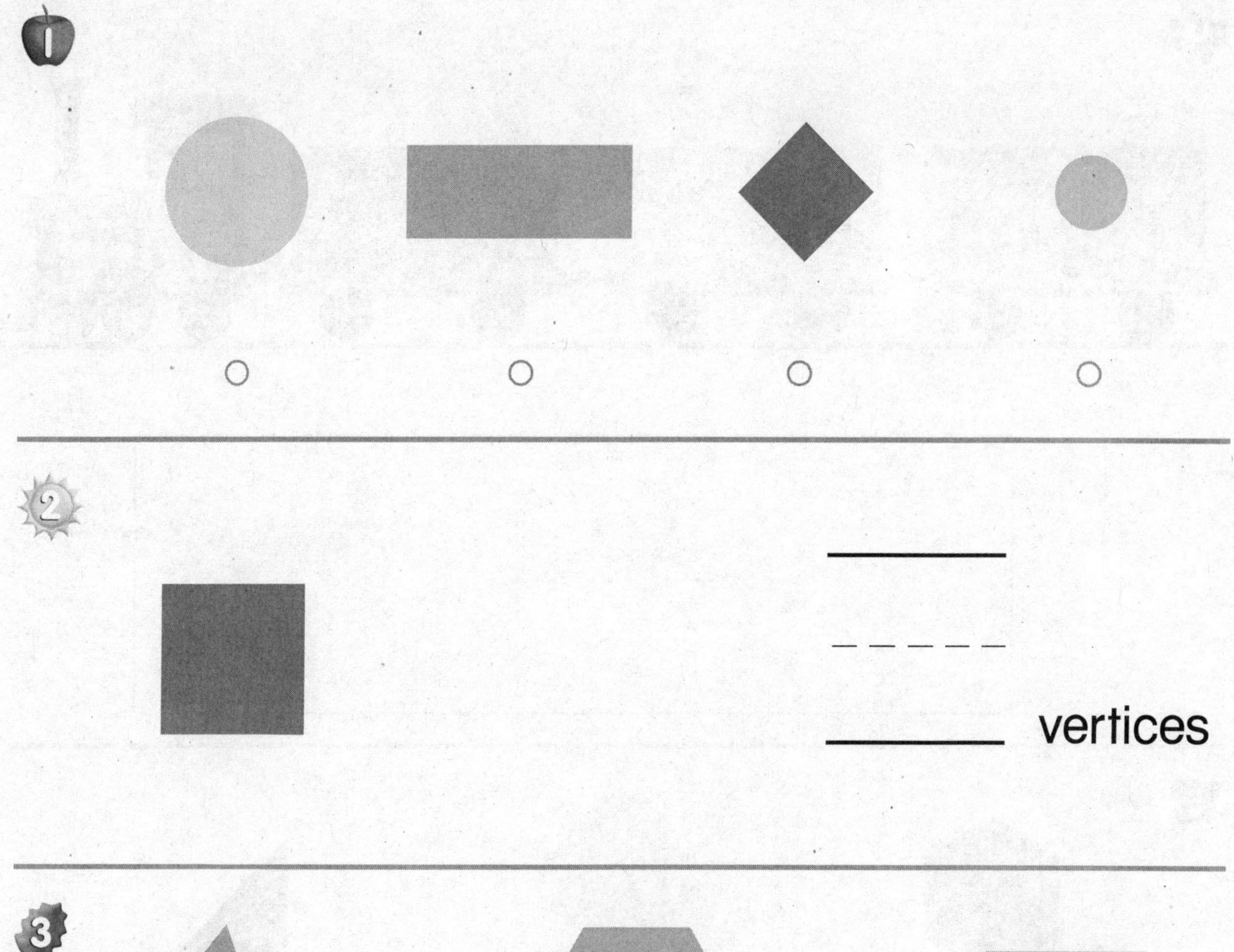

4 sides 3 sides 6 sides

DIRECTIONS **1.** Mark under all of the shapes that have curves. **2.** Look at the square. Write the number of corners, or vertices, the square has. **3.** Match the shape to the number of sides.

GO ON

Name ______________________________

4

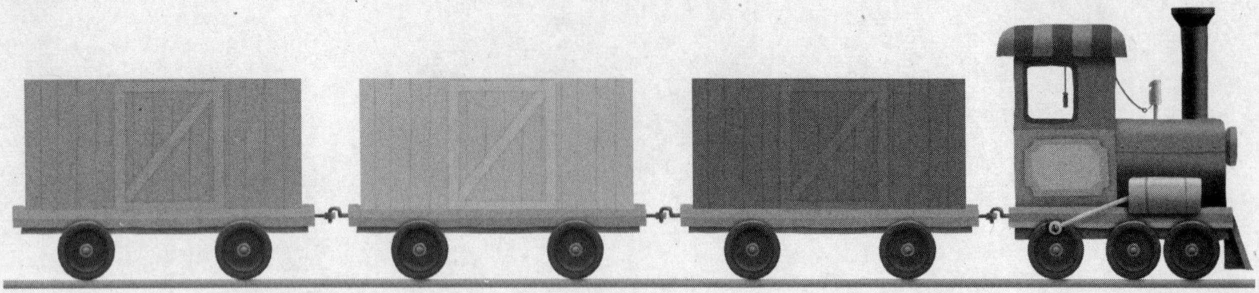

5

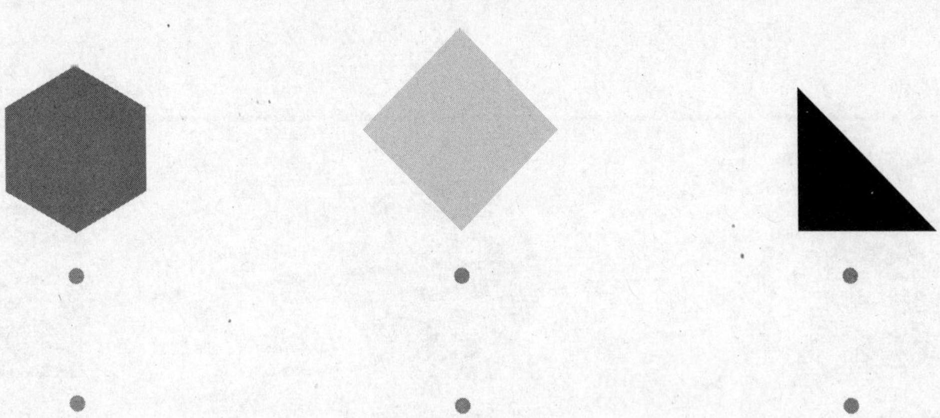

4 vertices　　3 vertices　　6 vertices

DIRECTIONS **4.** Draw a shape that is the same as the boxcars on the train. **5.** Match the shape to the number of corners, or vertices.

Name ______________________________

Practice Test
K.G.B.5
Analyze, compare, create, and compose shapes.

DIRECTIONS 1. Draw an object that has the shape of a cone. 2. Draw a triangle.

Name ______________________________

DIRECTIONS **3.** Draw a square or an object that has the shape of a square. **4.** Draw an object that has the shape of a cylinder.

Name ______________________________

Practice Test
K.G.B.6
Analyze, compare, create, and compose shapes.

DIRECTIONS **1.** Draw the two shapes used to make the shape shown. **2.** Draw the two shapes used to make the arrow. **3.** Two triangles are joined. Draw a shape they could make.

GO ON

Name ______________________

DIRECTIONS **4–5.** Draw the two shapes used to make the shape shown. **6.** Two rectangles are joined. Draw a shape they could make.

Name ______________________

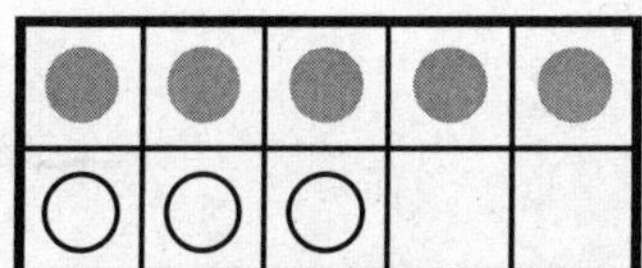

- ○ 5 plus 1
- ○ 5 plus 3
- ○ 5 + 3

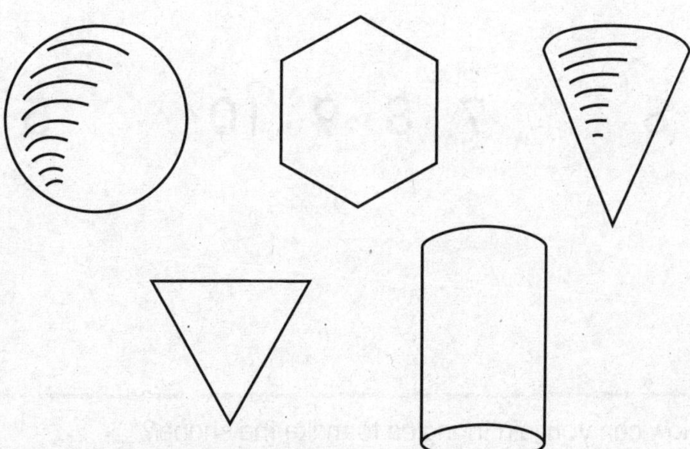

DIRECTIONS **1.** Circle the shapes that show the cone above the cube. **2.** Keri put 5 gray counters in the ten frame. Then she put 3 white counters in the ten frame. Choose all the ways that show the counters being put together. **3.** Color the solid shapes blue. Color the flat shapes red. Draw another flat shape that is different.

5

______

______

6

5 7 6 8 ○

7 8 9 10 ○

5 6 7 8 ○

DIRECTIONS 4. How can you join triangles to make the shape? Draw the triangles. 5. Count to tell how many. Write the number. 6. Mark under all sets of numbers that are in counting order.

GO ON

7

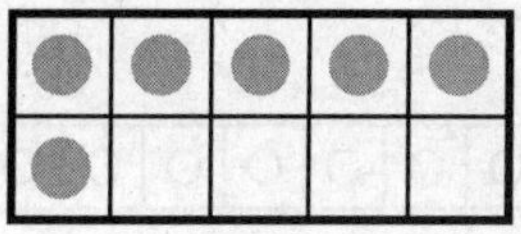

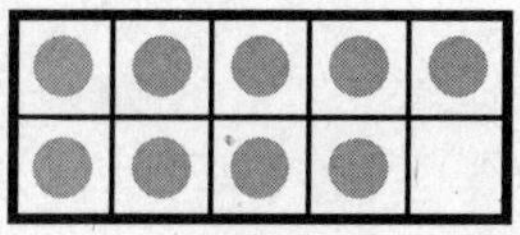

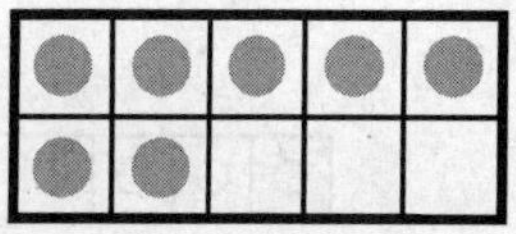

9 7 6

8

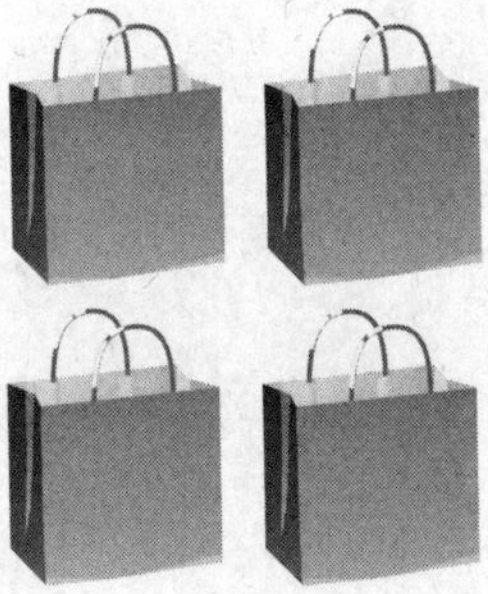

9

DIRECTIONS **7.** Match each set to the number that tells how many. **8.** Count each set of bags. Circle all the sets that show 3 bags. **9.** Draw 4 apples.

______ ______

8 ○ 5 ○ 6 ○

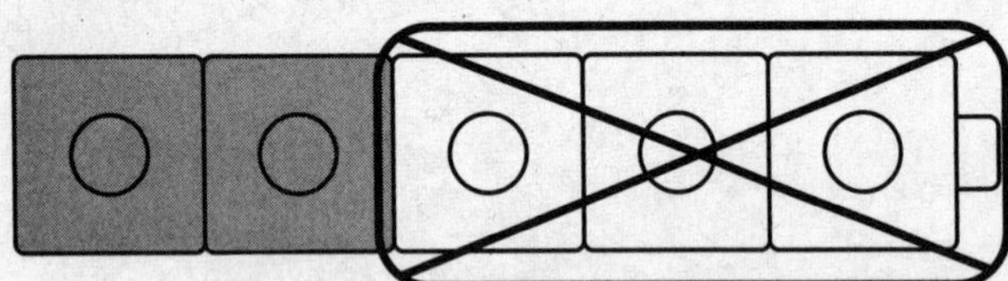

○ 5 − 4 = 1

○ 1 + 4 = 5

○ 5 − 3 = 2

DIRECTIONS **10.** Count how many in each set. Write the numbers. Circle the greater number. **11.** Think about counting order. Choose the number that is more than 7. **12.** Choose the number sentence that matches the picture.

GO ON

Name ______________________________

8 + 0   6 + 1   5 + 3

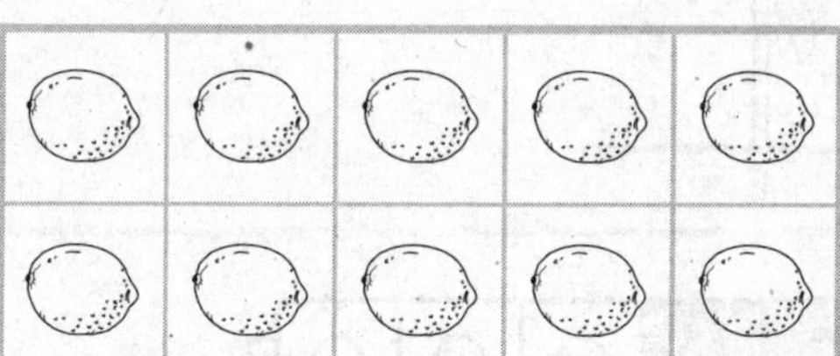
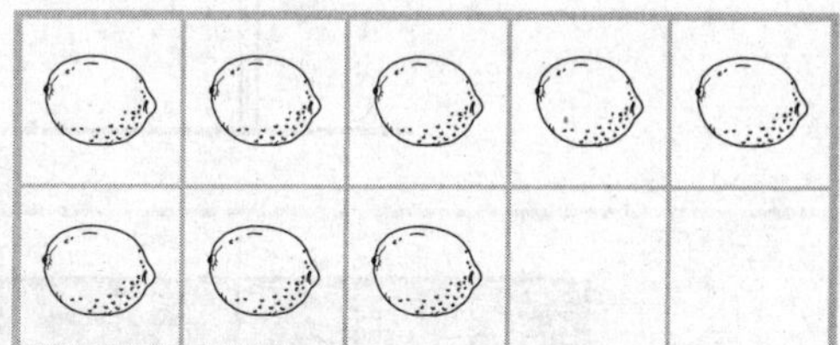

10 ones and [8 / 9] ones

DIRECTIONS **13.** Circle all the number pairs for 8. **14.** How many more ones are needed to show the number of lemons? Circle the number. **15.** Draw a line to show the height of the picture frame. Draw a line to show the length of the crayon.

GO ON

Name ______________________________

16

17

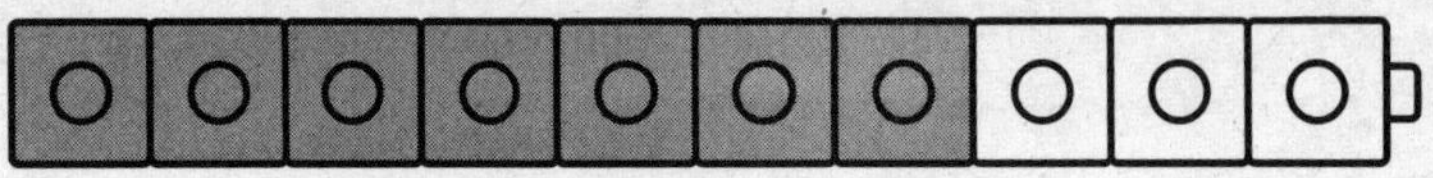

______ cubes

18

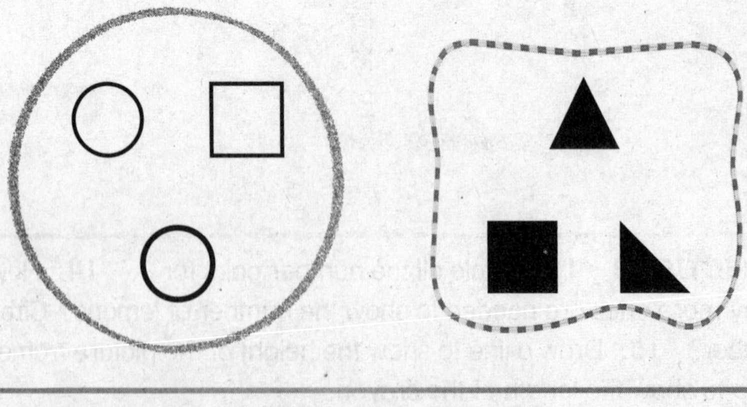

DIRECTIONS **16.** Circle the flower that is shorter. **17.** Write how many gray cubes. Write how many white cubes. Write how many cubes in all. **18.** Hani sorted some shapes into categories by color. Look at the shape at the beginning of the row. Mark an X on the category that shows where the shape belongs.

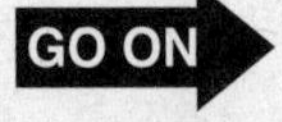

Name ____________________

DIRECTIONS **19.** Write the number that comes after 2 in counting order. Draw counters to show the number. **20.** Draw an object that has the shape of a sphere. **21.** Write and trace to complete the addition sentence.

GO ON

22

23

24

30 ______ 50 60 70

DIRECTIONS **22.** Mark an X on each shape that has 3 sides and 3 vertices. **23.** Circle the shapes that are circles. **24.** Count by tens. Write the missing number.

How Many Marbles?

2

____ − ____ = ____

DIRECTIONS **1.** Draw 8 marbles. Write the number of marbles you drew. **2.** Cross out 3 or 4 marbles. Write an equation to show how many marbles are left.

GO ON

Name ___________________________

10 = ___ + ___

10 = ___ + ___

DIRECTIONS Use counters. **3.** Bo has 7 blue marbles. Then he gets some red marbles. Now he has 10 marbles in all. Draw Bo's marbles. Write an equation to tell about Bo's marbles. **4.** Mia has 10 marbles. Four of her marbles are yellow and the rest are green. Draw Mia's marbles. Write an equation to tell about Mia's marbles.

GO ON

Name ______________________

5

6

DIRECTIONS **5.** Rory has 16, 17, or 18 marbles. Write a number that could be Rory's marbles. **6.** Place counters in the ten frames to show that number. Draw the counters. **7.** Complete the equation to show how to make that number.

Name ____________________

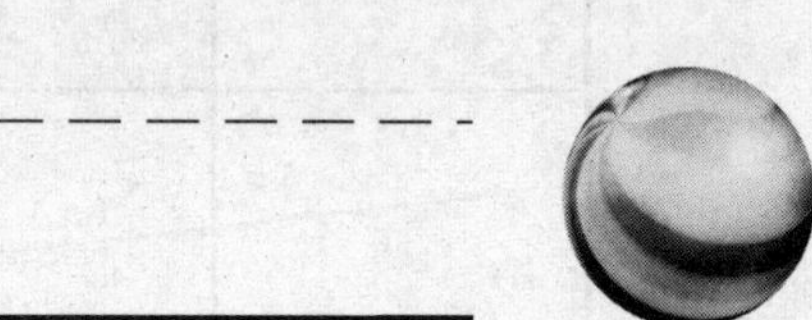

DIRECTIONS **8.** There are 6 bags of marbles on the table. Each bag has 10 marbles. Draw the bags of marbles. Count by tens to show how many marbles in all. Write the number. **9.** Sam has 14 marbles. Liam has 2 more marbles than Sam. Draw both sets of marbles. Circle the set that has a greater number of marbles.

Name ____________________

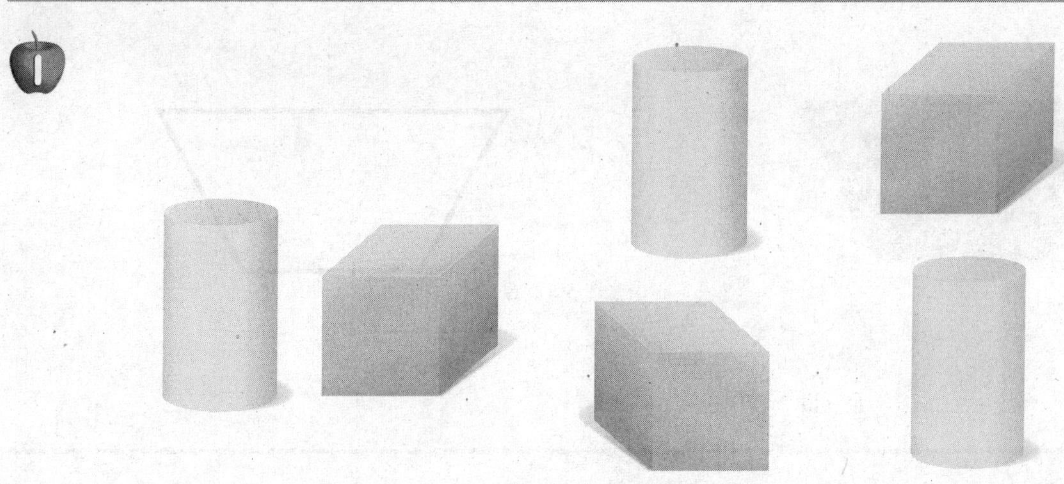

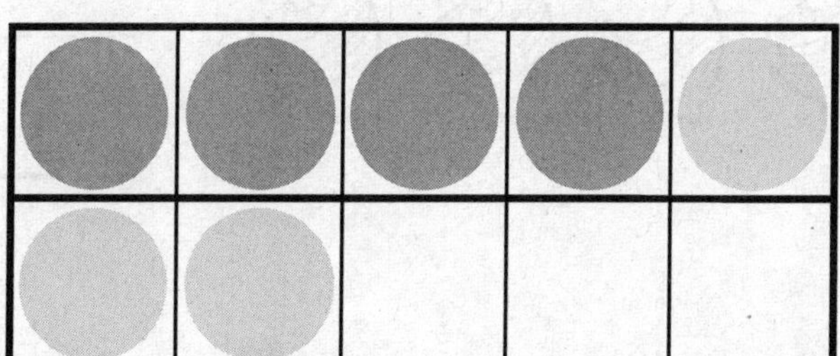

- ○ 4 plus 3
- ○ 4 plus 1
- ○ 4 + 1

DIRECTIONS 1. Circle the shapes that show the cylinder above the cube. **2.** Sonja put 4 dark gray counters in the ten frame. Then she put 3 light gray counters in the ten frame. Choose all the ways that show the counters being put together. **3.** Color the solid shapes blue. Color the flat shapes red. Draw another flat shape that is different.

GO ON

5

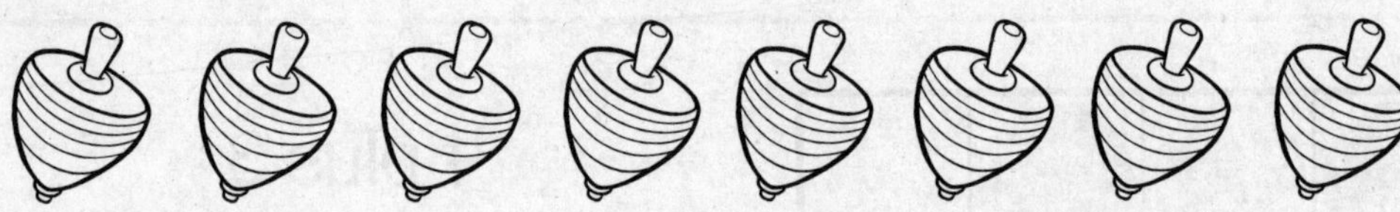

6

5 6 7 8 ○

8 10 9 7 ○

7 8 9 10 ○

DIRECTIONS **4.** How can you join triangles to make the shape? Draw the triangles. **5.** Count to tell how many. Write the number. **6.** Mark under all sets of numbers that are in counting order.

Name ____________________________________

7

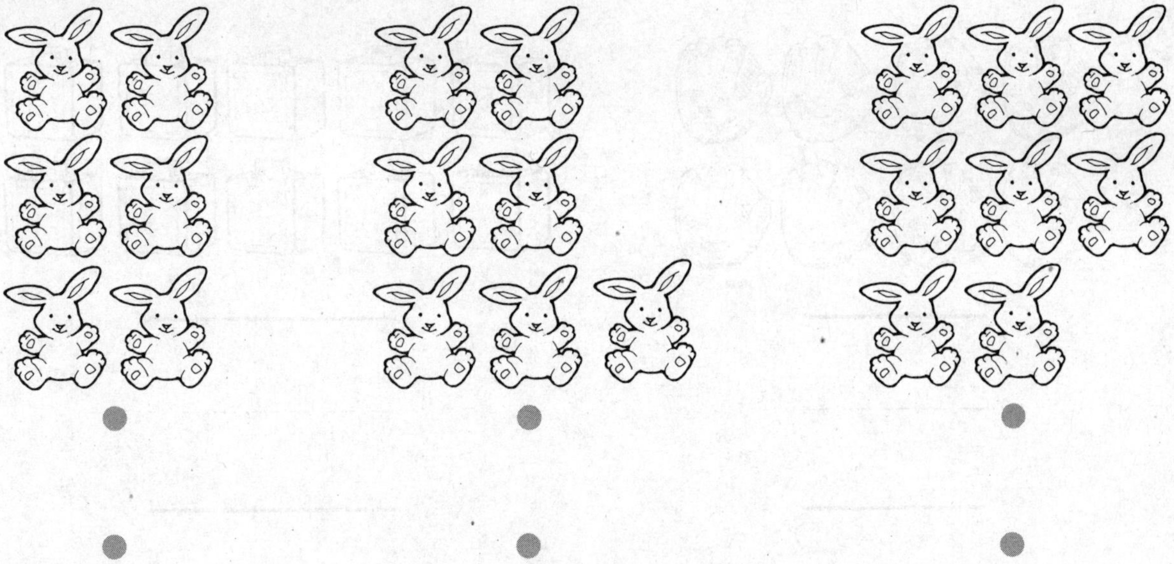

8 6 7

8

9

DIRECTIONS 7. Match each set to the number that tells how many. 8. Count each set of bags. Circle all the sets that show 2 bags. 9. Draw 2 fish.

10

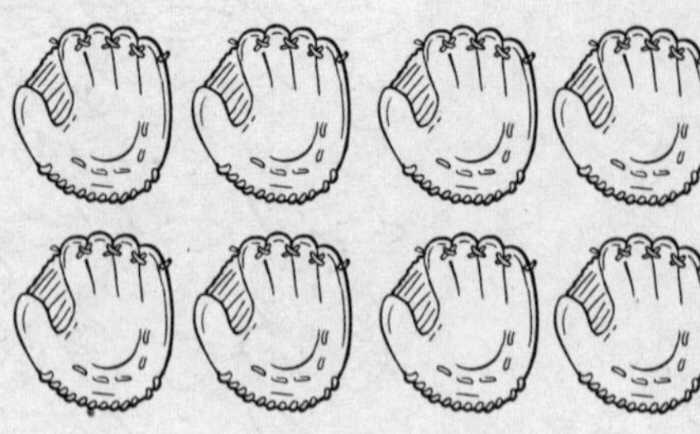

11

7 10 9

○ ○ ○

12

○ 5 − 4 = 1

○ 4 + 1 = 5

○ 5 − 2 = 3

DIRECTIONS 10. Count how many in each set. Write the numbers. Circle the greater number. 11. Think about counting order. Choose the number that is less than 8. 12. Choose the number sentence that matches the picture.

GO ON

Name ______________________

4 + 5 2 + 6 1 + 7

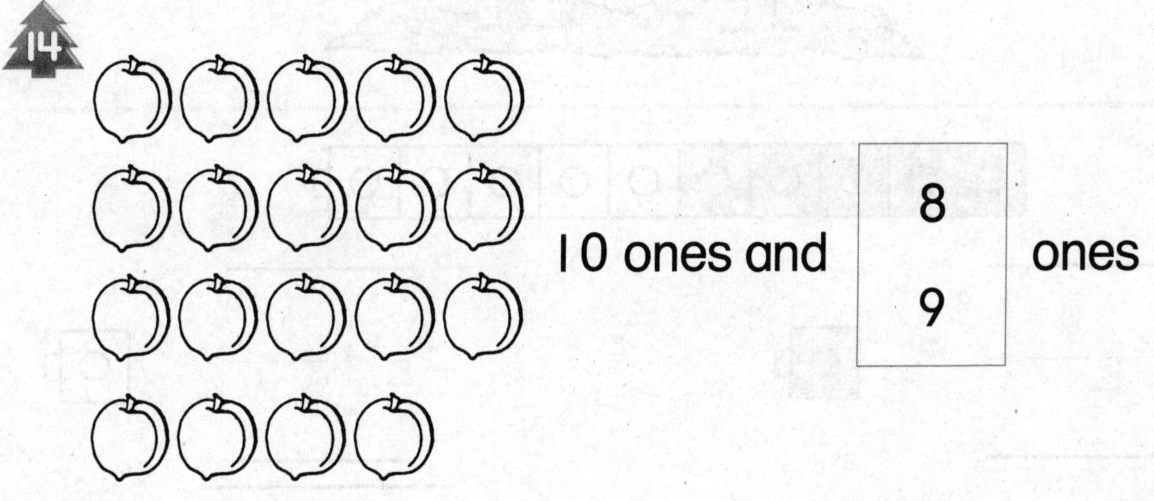

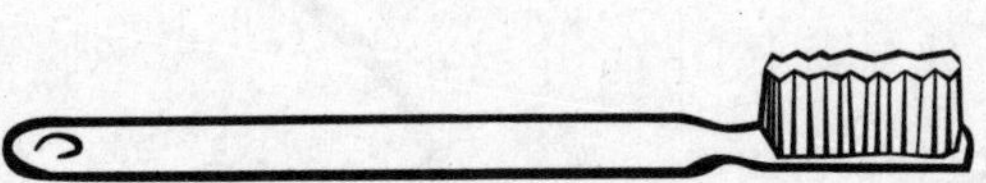

DIRECTIONS **13.** Circle all the number pairs for 8. **14.** How many more ones are needed to show the number of peaches? Circle the number. **15.** Draw a line to show the height of the lamp. Draw a line to show the length of the toothbrush.

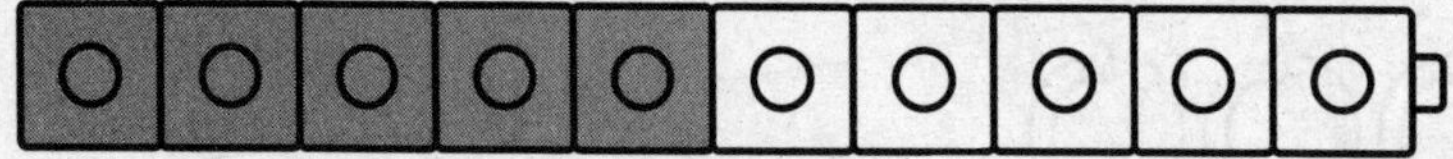

______

______

______

______

______

______ cubes

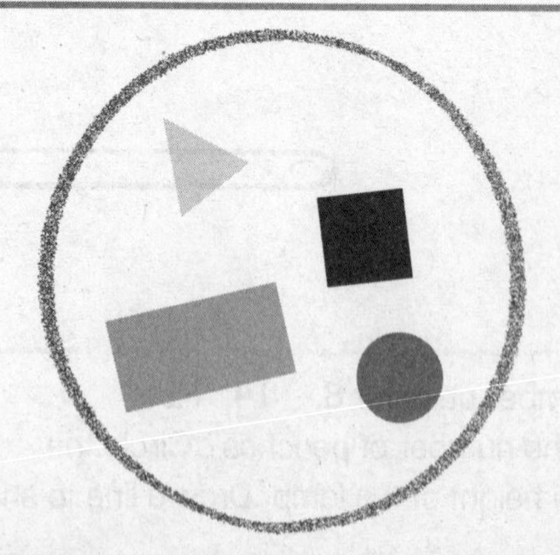

DIRECTIONS **16.** Circle the tree that is taller. **17.** Write how many gray cubes. Write how many white cubes. Write how many cubes in all. **18.** Lin sorted some shapes into categories by color. Look at the shape at the beginning of the row. Mark an X on the category that shows where the shape belongs.

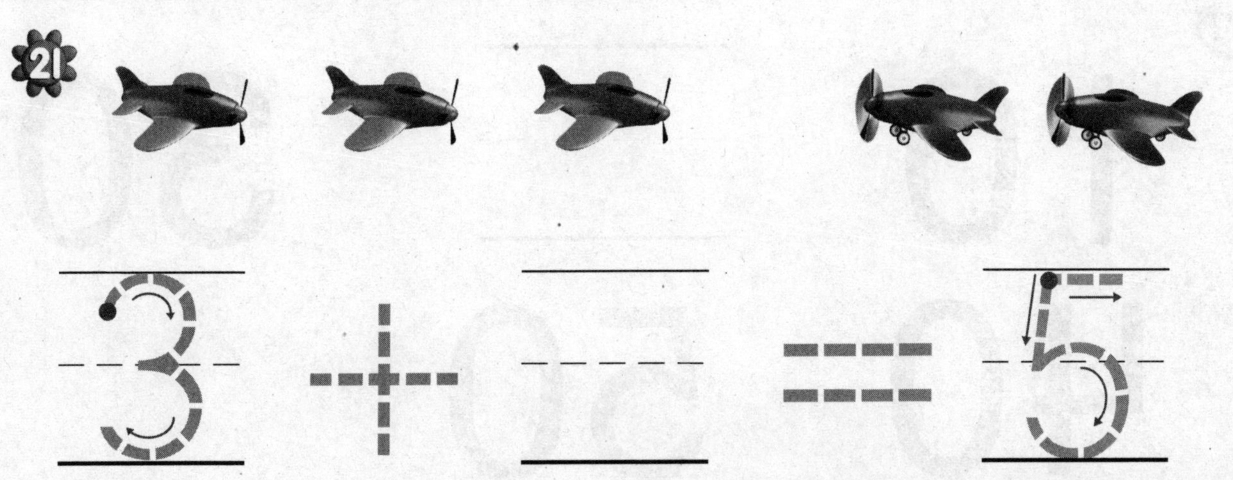

DIRECTIONS **19.** Write the number that comes after 1 in counting order. Draw counters to show the number. **20.** Draw a rectangle. **21.** Write and trace the numbers to complete the addition sentence.

23

24

10 ___ 30

40 50

DIRECTIONS **22.** Mark an X on each shape that has 3 sides and 3 vertices. **23.** Circle the shapes that are circles. **24.** Count by tens. Write the missing number.

Name ______________________

Middle-of-Year
Performance Task

Shapes, Shapes, Shapes!

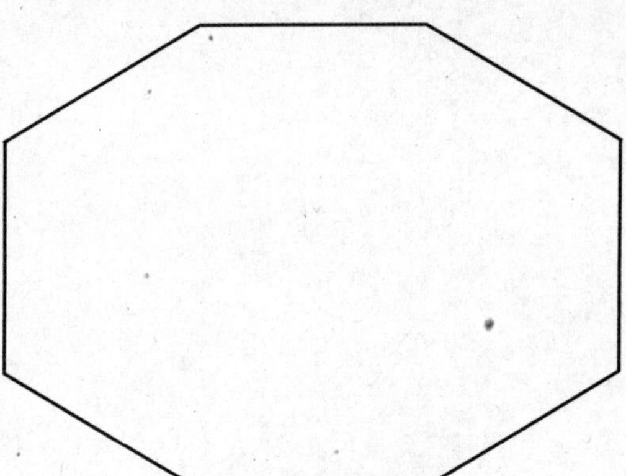

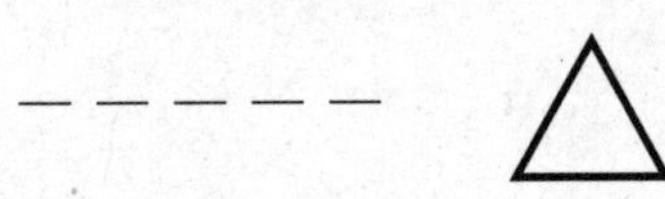

DIRECTIONS **1.** Use triangles and squares to create the big shape. Draw the shapes you used. **2.** How many triangles did you draw? Write the number. **3.** How many squares did you draw? Write the number.

GO ON

DIRECTIONS Use your shapes. **4.** Trace the rectangle. Color it green. Above the green shape, draw a shape with no sides and no vertices. Color this shape yellow. Under the green shape, draw a shape that has 3 sides and 3 vertices. Color this shape blue. Next to the blue shape, draw a shape that has 6 sides and 6 vertices. Color this shape red. Put an X on any shape that has more than 3 sides.

GO ON

Name ____________________

5

DIRECTIONS 5. Find all the shapes that have no flat surfaces and can roll. Color them red. Find all the shapes that have six flat surfaces and can stack. Color them blue. Find all the shapes that have one flat surface and can roll. Color them green. Find all the shapes that have two flat surfaces and can roll. Color them yellow.

GO ON

Name ________________________________

Middle-of-Year Performance Task

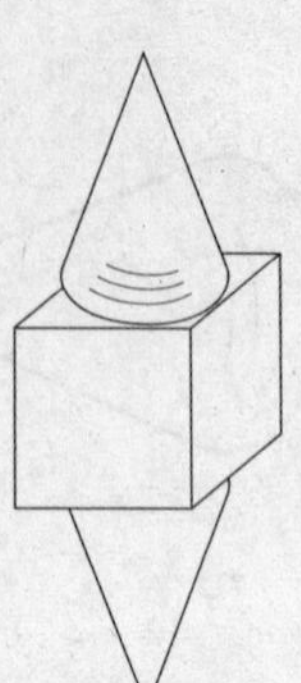

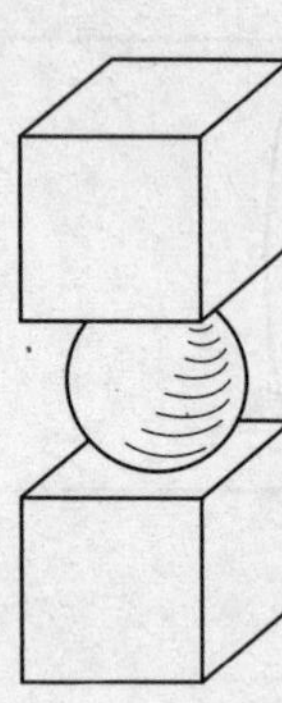

DIRECTIONS **6.** John is moving to a new house. He wants to stack items on a moving truck. Circle the picture that shows a way John could stack his items. Then find a picture with cubes that are above and below a sphere. Color that picture. **7.** Draw a way that John could stack a cube and a cone. Tell or write how you stacked the shapes. **8.** Draw a shape that can roll next to a shape that cannot roll. Write the names of the shapes you drew.

Name ______________________________

1

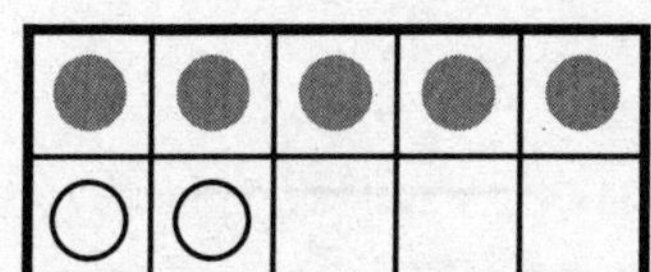

- ○ 5 plus 1
- ○ 5 plus 2
- ○ 5 + 2

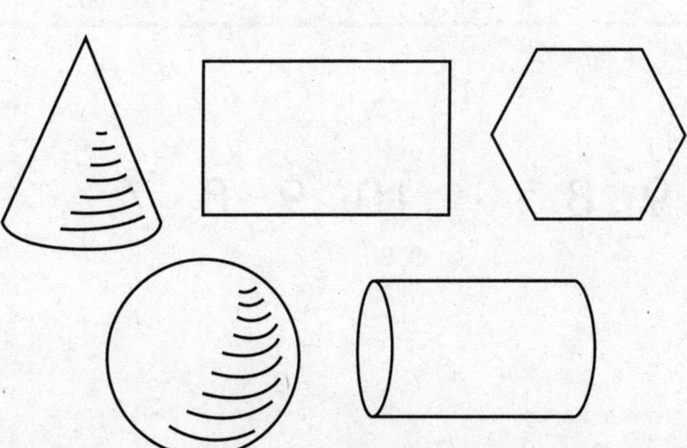

DIRECTIONS 1. Circle the shapes that show the cylinder below the cube. **2.** Jake put 5 gray counters in the ten frame. Then he put 2 white counters in the ten frame. Choose all the ways that show the counters being put together. **3.** Color the solid shapes blue. Color the flat shapes red. Draw another flat shape that is different.

GO ON

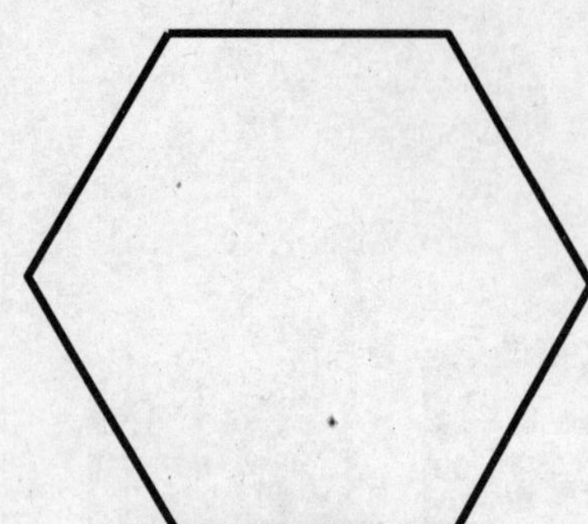

5

6

6 7 9 8 ○

10 9 8 7 ○

6 7 8 9 ○

DIRECTIONS 4. How can you join triangles to make the shape? Draw the triangles. 5. Count to tell how many. Write the number. 6. Mark under all sets of numbers that are in counting order.

GO ON

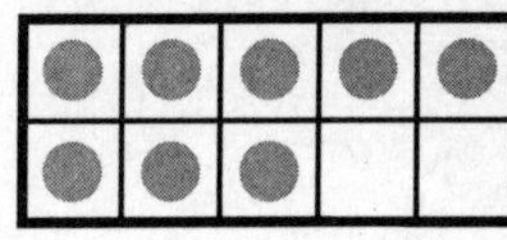

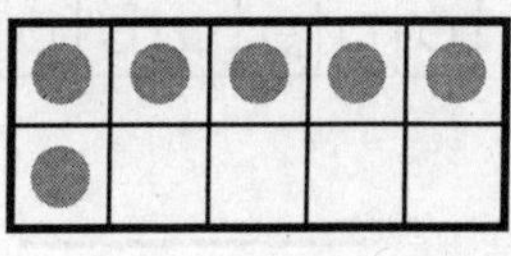

8 6 5

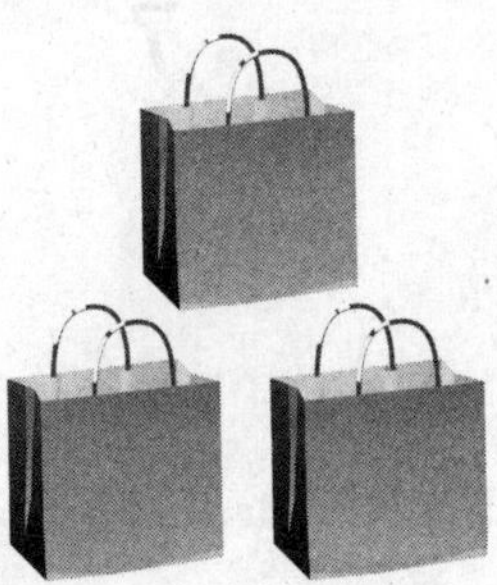

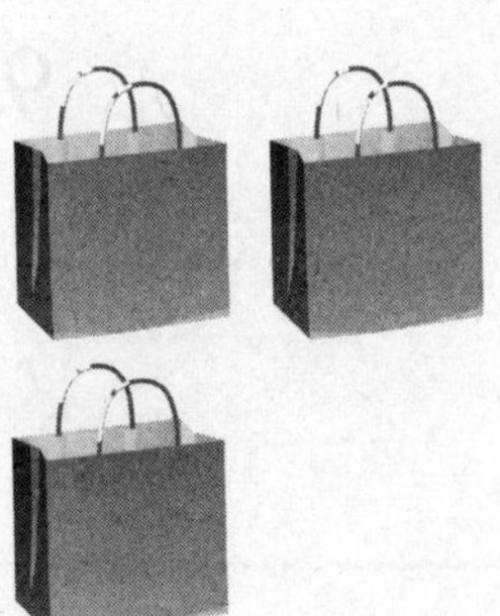

DIRECTIONS **7.** Match each set to the number that tells how many. **8.** Count each set of bags. Circle all the sets that show 3 bags. **9.** Draw 3 flowers.

10

______

______

______

______

11

6 ○ 9 ○ 7 ○

12

○ $5 - 1 = 4$

○ $1 + 4 = 5$

○ $5 - 3 = 2$

DIRECTIONS **10.** Count how many in each set. Write the numbers. Circle the greater number. **11.** Think about the counting order. Choose the number that is more than 8. **12.** Choose the number sentence that matches the picture.

GO ON

Name ______________________________

6 + 2   1 + 7   4 + 5

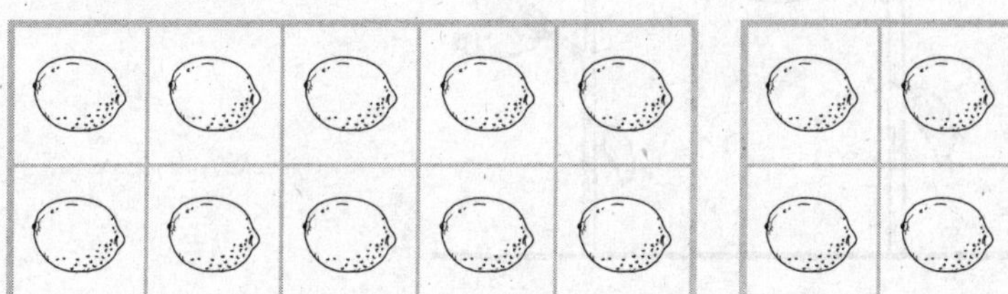

10 ones and | 8 / 9 | ones

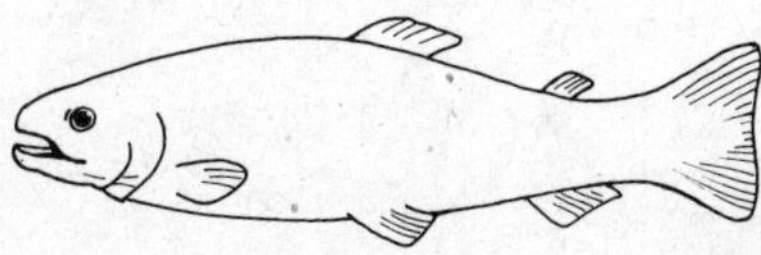

DIRECTIONS 13. Circle all the number pairs for 8. 14. How many more ones are needed to show the number of lemons? Circle the number. 15. Draw a line to show the height of the jar. Draw a line to show the length of the fish.

GO ON

Name ______________________

17

cubes

18

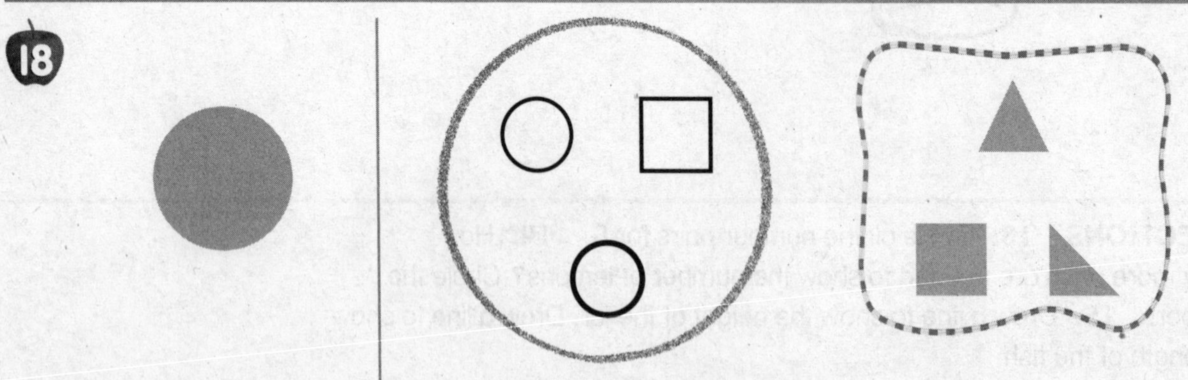

DIRECTIONS **16.** Circle the flower that is taller. **17.** Write how many gray cubes. Write how many white cubes. Write how many cubes in all. **18.** Pablo sorted some shapes into categories by color. Look at the shape at the beginning of the row. Mark an X on the category that shows where the shape belongs.

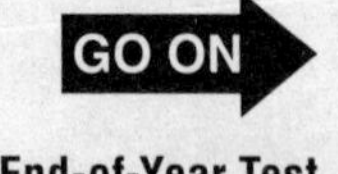

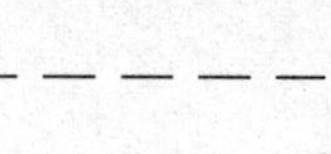

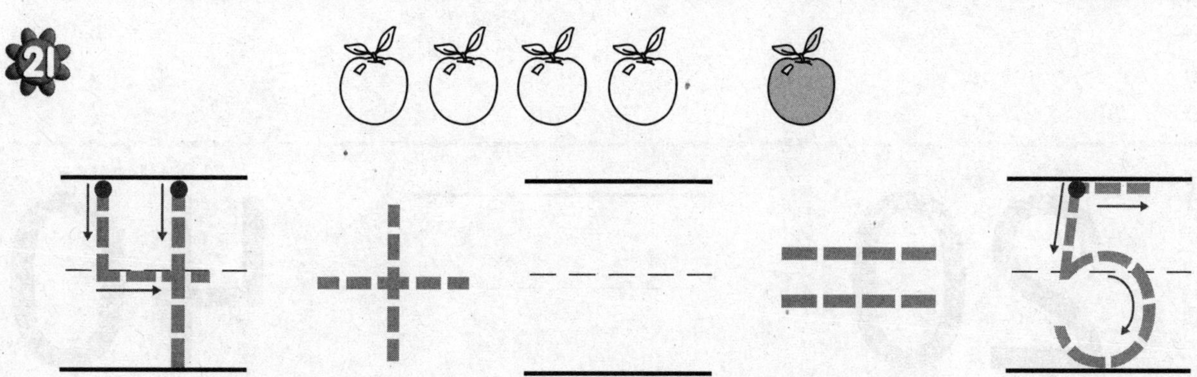

DIRECTIONS **19.** Write the number that comes after 3 in counting order. Draw counters to show the number. **20.** Draw an object that has the shape of a cylinder. **21.** Write and trace to complete the addition sentence.

22

24

20 ______ 40

50 60

DIRECTIONS **22.** Mark an X on each shape that has 4 sides and 4 vertices. **23.** Circle the shapes that are squares. **24.** Count by tens. Write the missing number.

Name ______________________

Comparing and Sorting

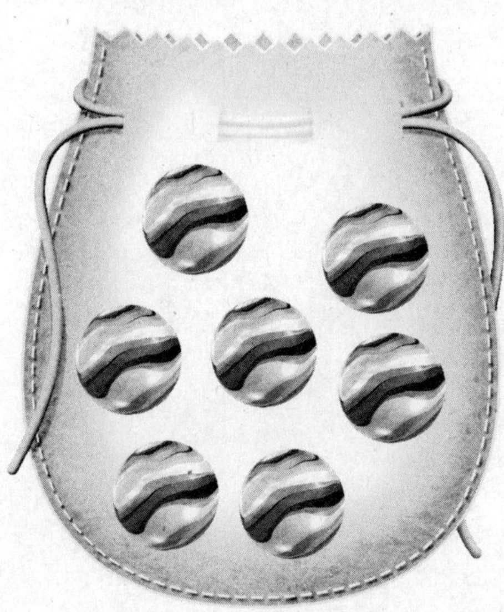

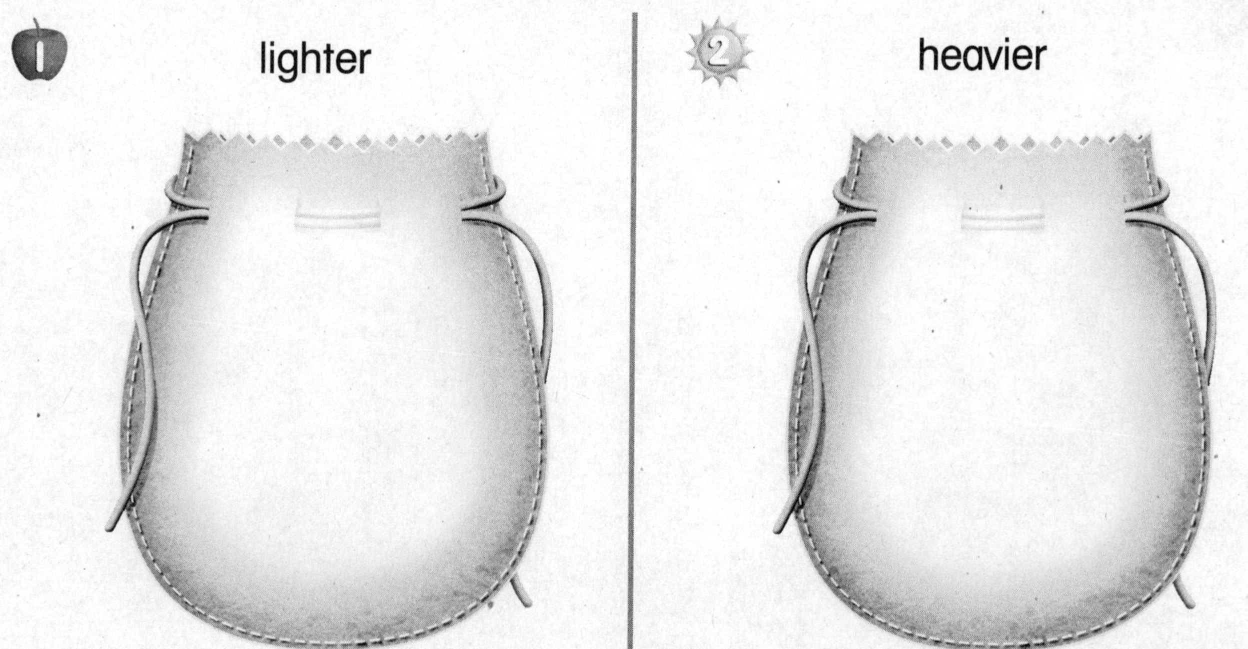

DIRECTIONS **1.** Felipe has a bag of marbles. Draw marbles in the first bag to make a bag that is lighter than Felipe's bag. **2.** Draw marbles in the second bag to make a bag that is heavier than Felipe's bag.

Name ______________________________

DIRECTIONS **3.** Draw two worms. Make one longer than the other. Circle the worm that is shorter. **4.** Draw two trees. Make one tree shorter than the other. Circle the tree that is taller.

5

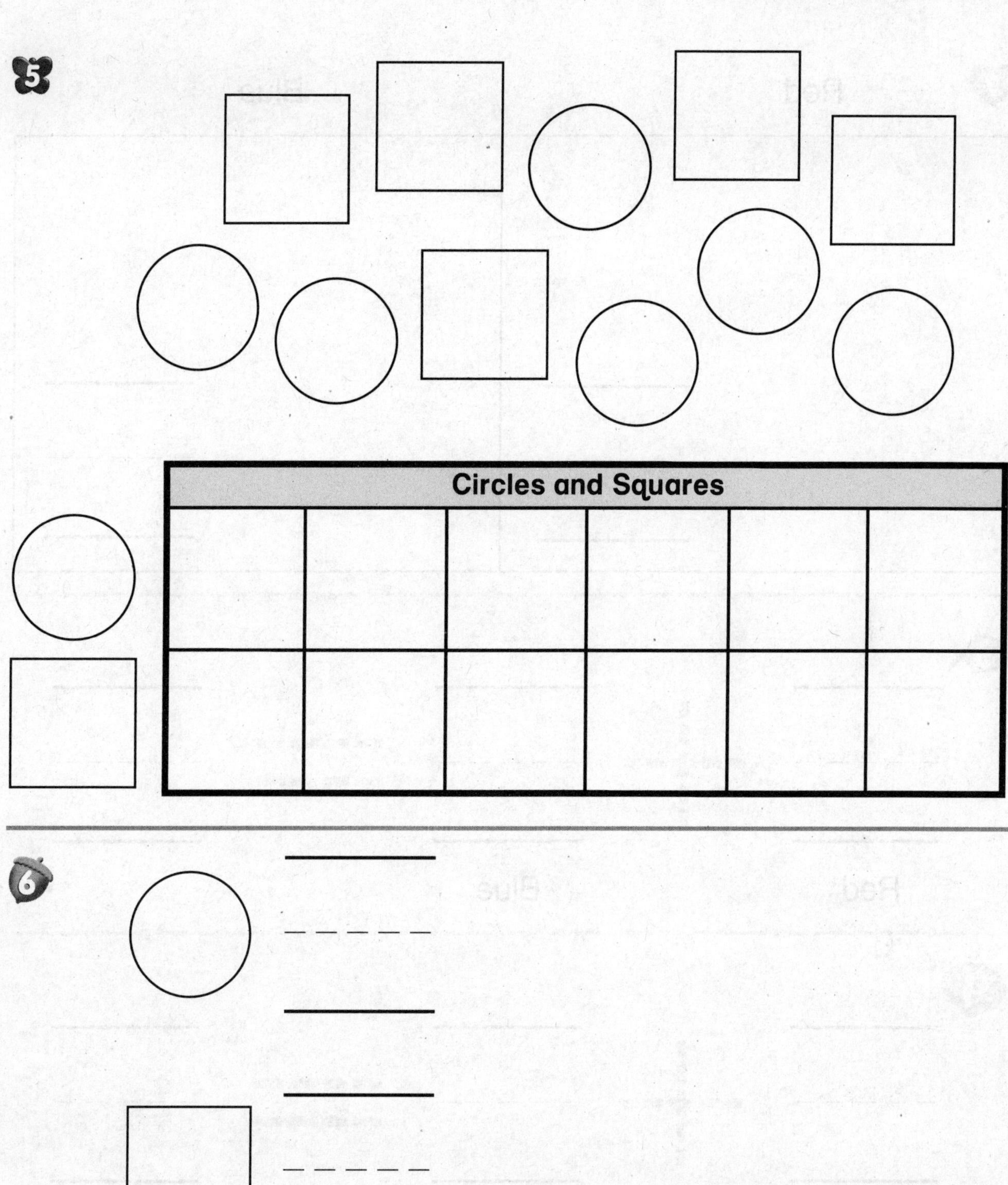

6

DIRECTIONS 5. Look at the circles and squares. Sort the shapes to complete the graph. Use your shape blocks to help you. 6. Write how many of each.

GO ON

Name ______________________________

7 Red Blue

8 ______ + ______ = ______

Red Blue

9 ______ + ______ = ______

DIRECTIONS 7. Draw 1, 2, or 3 red shapes in the first box. Write the number of red shapes you drew. Draw 4, 5, or 6 blue shapes in the second box. Write the number of blue shapes you drew. 8. Write the number sentence for the number of shapes in all. 9. Write the number sentence a different way.

Name ______________________

Year-End Performance Assessment Task

Ella's Art Supplies

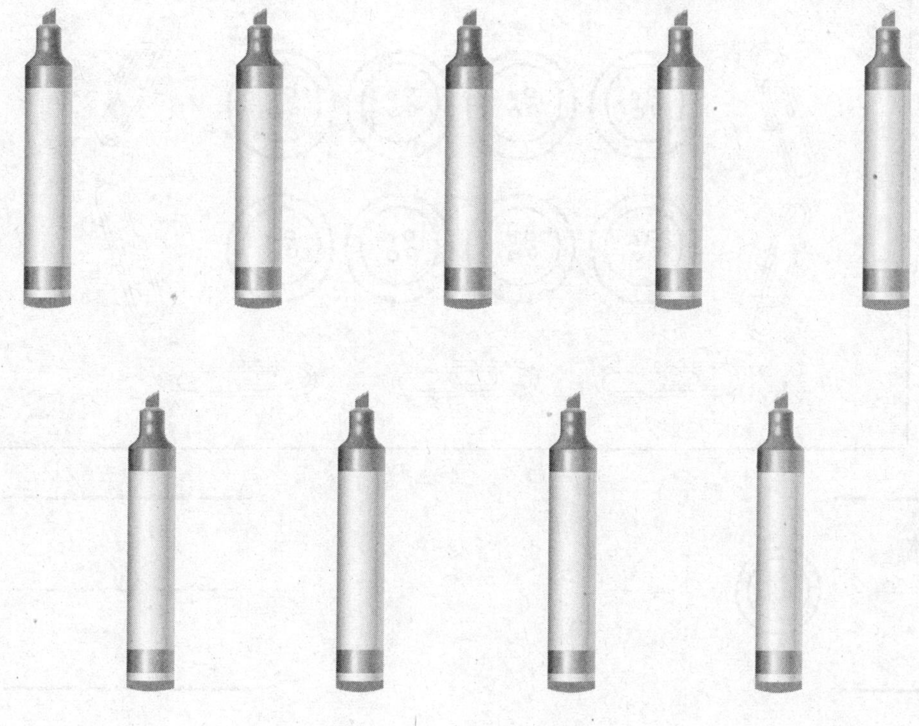

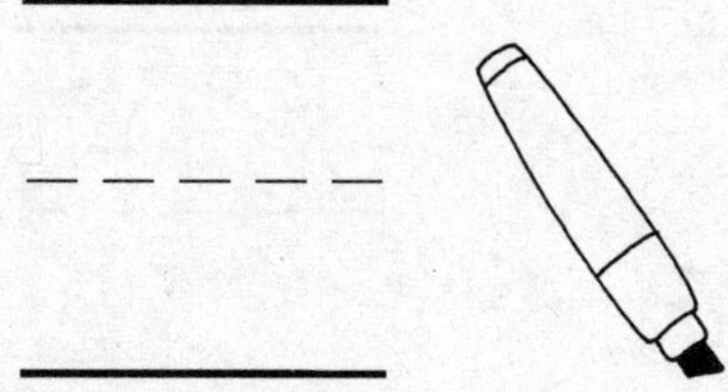

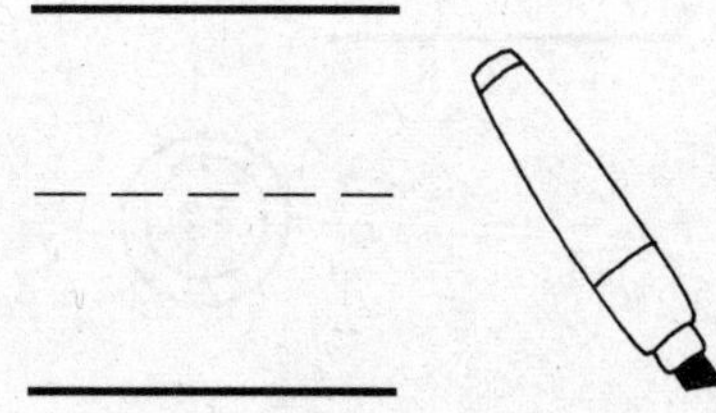

red

blue

DIRECTIONS **1.** Ella has 9 markers. Some are red and some are blue. Use connecting cubes to show a number of red and blue markers Ella could have. Color the markers. Then write the numbers of red and blue markers.

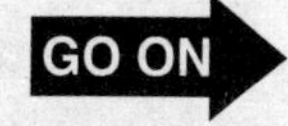

Name ______________________

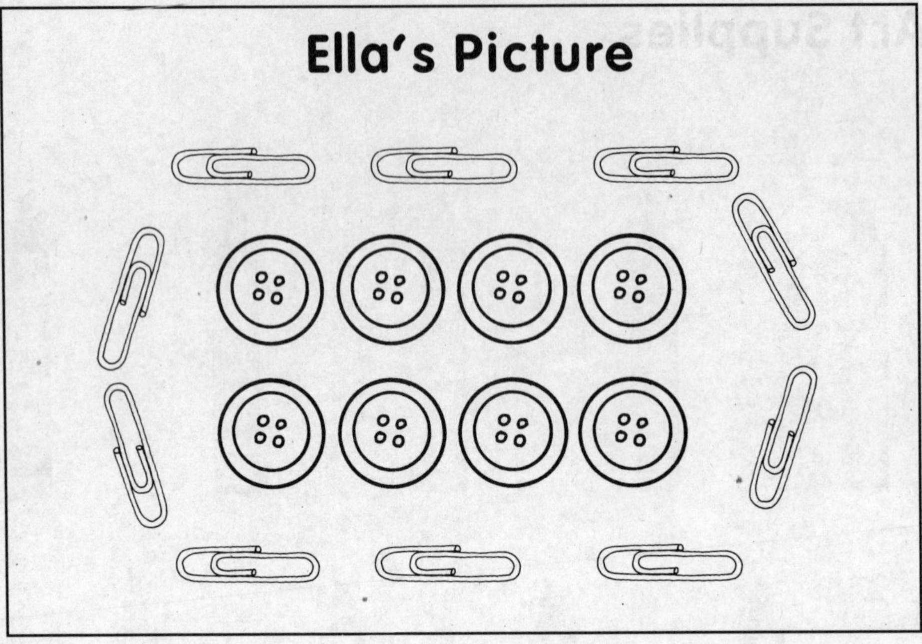

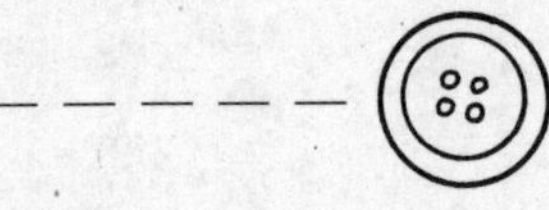

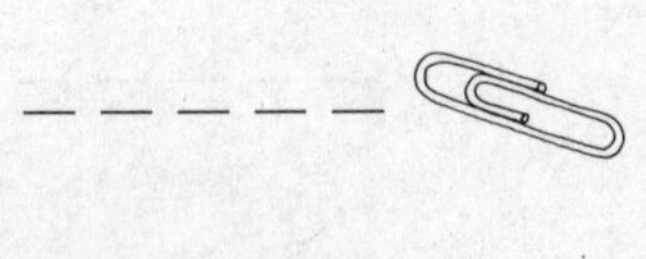

DIRECTIONS **2.** Ella created a picture using buttons and paper clips. How many buttons did Ella use? Write the number. How many paper clips did Ella use? Write the number. Compare the numbers. Circle the greater number. **3.** Draw your own picture with some number of buttons less than 10. Draw a number of paper clips that is 1 less than the number of buttons. Write the number of buttons and paper clips you drew. Compare the numbers. Circle the greater number.

5 – ____ = ____

5

5 – ____ = ____

DIRECTIONS 4. Ella has 5 stickers. She gives some to Ben. Mark an X on each sticker you want Ella to give to Ben. How many stickers does Ella have left? Write numbers in the sentence to show what you did. 5. Mark a different number of stickers for Ella to give to Ben. Complete the number sentence.

GO ON

Name ______________________________

10 = ____ + ____

10 = ____ + ____

10 = ____ + ____

10 = ____ + ____

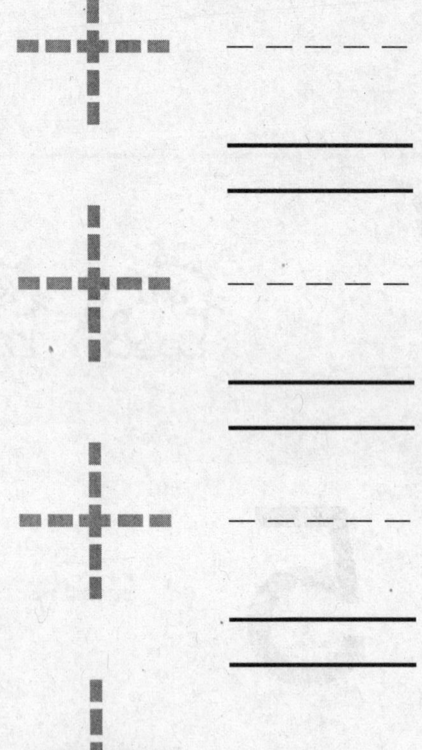

DIRECTIONS **6.** Ella has 10 crayons. She puts the crayons in two boxes. Use counters to show four different ways Ella could put the crayons in the boxes. Write the addition sentence for each way. Then circle one of your addition sentences. Draw crayons in the box to match your addition sentence.